D1014888

The
Quest for
Unity

The Quest for Unity

• • •

The Adventure of Physics

Étienne Klein and Marc Lachièze-Rey
TRANSLATED BY AXEL REISINGER

New York Oxford
Oxford University Press
1999

Oxford University Press

Oxford New York
Athens Auckland Bangkok Bogotá Buenos Aires Calcutta
Cape Town Chennai Dar es Salaam Delhi Florence Hong Kong Istanbul
Karachi Kuala Lumpur Madrid Melbourne Mexico City Mumbai
Nairobi Paris São Paulo Singapore Taipei Tokyo Toronto Warsaw

and associated companies in
Berlin Ibadan

Originally published by Editions Albin Michel,
22, rue Huyghens, 75680 Paris Cedex 14, France
Original French title: La Quête de l'Unité: L'Aventure de la Physique

Published by Oxford University Press, Inc.
198 Madison Avenue, New York, New York 10016

Library of Congress Cataloging-in-Publication Data
Klein, Étienne.
(Quête de l'unité)
The quest for unity: the adventure of physics / by Étienne Klein and Marc Lachièze-Rey;
translated by Axel Reisinger.
p. cm. Includes index.
ISBN 0-19-512085-X
1. Physics—History.
2. Grand unified theories (Nuclear physics)
I. Lachièze-Rey, Marc.
II. Title.
QC7.K5413 1999
530'.09—dc21 98-29622

Oxford University Press would like to acknowledge
a generous subsidy from the French Ministry of Culture
toward the cost of translating this work into English.

9 8 7 6 5 4 3 2 1
Printed in the United States of America
on acid-free paper

100

Contents

Introduction

The number one shall be defined in Chapter 3.
—Nicolas Bourbaki

The immenseness and diversity of the universe are awe-inspiring. The richness of the creation and the myriad forms it assumes seem far too boundless for us ever to hope to encapsulate them in a few principles. Can the world even be fathomed? The very idea of an underlying unity seems preposterous, almost insane right from the outset. The moment it is articulated, it is belied by the world itself. What could a star, a cloud, a snowflake, a living cell, an atom, and a quark possibly have in common? How could anyone reconcile the notion of unity with so many plausible and reasonable arguments that justify splintering reality into disconnected pieces? Indeed, everything points to multiplicity as an unassailable characteristic of the physical world.

And yet, without unity as a beacon, the world, indeed human thought itself, would scatter into a dust of things and ideas impossible to integrate. The very concept of universe would become senseless. The history of human thought offers a decidedly different panorama, one that is full of syntheses, bridges, unifications, and sometimes even outright fusions. The most remarkable successes have revolved around matter and its interactions. The credit goes to men of science who have managed to distill from a profusion of phenomena and concepts a few guiding principles capable of organizing and unifying the infinite diversity of everyday experience. Indeed, almost every single "great physicist" recorded by history has contributed unifications that changed the reach and power of physics, transforming it into something more than a simple patchwork of disparate theories.

Galileo, for instance, reconciled the sublunar and supralunar worlds when his telescope revealed mountains and valleys on the surface of the moon. Newton created a single theory to describe the motion of the earth and that of the celestial bodies. Maxwell unified electricity and magnetism. Fraunhofer demonstrated that the physical laws discovered here on earth apply just as well to stars. Louis de Broglie established a connection between waves and particles. Perhaps the most famous case is that of Albert Einstein, who melded space and time, until then completely distinct, into a unified space-time concept. He later devoted much energy looking for a unifying theory that would encompass both the universe taken as a whole and the laws governing its elementary constituents.

But from this vantage point, does science really differ from thought itself? The longing for unity is without a doubt a requisite for intelligibility. It is a fundamental need of man's intellect with its thirst for synthesis. At least, such was Leibniz's opinion. He felt that the One (together with being, substance, sameness, cause, perception, and reasoning) is part of the innate notions anchored in our minds, without which the data of human experience would remain incomprehensible.[1] Immanuel Kant shared the same view. He described understanding as "the power to reduce phenomena to the unity of rules," and reason as "the ability to subjugate the rules of understanding to unity by means of principles." It would be difficult to articulate more clearly the monism pervading the pursuit of knowledge.

Be that as it may, it is in the sciences that the quest for unity has produced the most spectacular results. It continues to stimulate a considerable amount of research. During the 1970s, two types of phenomena—electromagnetism and weak interactions (the latter being responsible for, among other things, the decay of a neutron into a proton, an electron, and a third particle called antineutrino)—ostensibly quite dissimilar in their phenomenology, were formally united within the framework of a new and broader theory. The so-called *standard model* was born, providing a very elegant classification scheme of the constituents of matter into three families. This encouraging success seemed to bring us one step closer to a triumphant unification.

By all evidence, the drive toward unity and synthesis is the primary impetus behind many scientific endeavors. It seems to have been a particularly fruitful and effective tool in the field of physics. Indeed, the belief that unification is the very foundation of physics and constitutes its ultimate mission is widespread. Pierre Duhem said as much explicitly: "Every physicist naturally aspires to the unity of science."[2] This aspiration is twofold: It reflects a drive "toward the logical unity of physical theory…[and] toward a theory which is a natural classification of physical laws."[3]

With the help of principles and laws applied to a great variety of phenomena, physics has allowed us to unravel the organization of things by resolving complex structures into more "fundamental" substructures. Proceeding down this line, matter proved to be made of molecules, molecules of atoms, atoms of nuclei and electrons, nuclei of quarks and gluons. Who knows where it might all end? In this light, physics appears to be inherently reductionistic. Many of its greatest accomplishments can be viewed as "reducing" or submitting theory A to theory B. All laws known on the basis of A can be inferred in terms of laws implied by B through a systematic process of logical deduction. Such connections have undoubtedly convinced many a physicist that hidden within the luxuriant abundance of the world lurk elementary objects and underlying theories apt to shatter the outer appearance of contingency and specificity.

The notion of unity has repeatedly demonstrated its relevance and power throughout the history of physics. Yet it is not easy to grasp its true nature. The very idea of unity remains controversial. There are those who will forever maintain their conviction (1) that the universe is inherently inextricable; (2) that its essence is the type of raw diversity some philosophers use to define matter, to the point that no law or federation could ever hold sway over it; and (3) that the very idea of unity is nonsense because any thought belongs inexorably in the realm of the multiple. They might even reach the conclusion that reason has developed an inordinate cult of unity, fostered by an equally unjustified disdain for multiplicity. After all, why should space-time and everything it contains be forced to yield to the overbearing grip of human thought?

Some outstanding scholars have in fact denounced the myth of unity. In the middle part of the nineteenth century, Antoine Augustin Cournot, for one, argued that the fundamental historicity of science makes reason incapable of enfolding the constitution and order of things into a single concept. In his view, the reason for this inability to truly unify knowledge is not methodological but ontological, because the very essence of the world is fundamentally diverse and segmented. Reality is made of things whose links are at times close, and at other times loose or even nonexistent. "It is not within the capability of an intelligence like ours, nor of any other finite intelligence, to capture in a single system the phenomena and laws of all of nature," Cournot wrote. "And even if we could, we would still identify in such a system parts that are disconnected and form the object of unrelated theories, even though they might well be traceable to a common origin."[4] In short, this position makes no allowance for any hierarchy of things or even for a universal set of experiences to which a single system of categories might give meaning and intelligibility.

There are other, no less brilliant scholars who have argued exactly the opposite. Any discipline or intellectual endeavor is, by its very nature, susceptible to what Blaise Pascal called in his *Pensées* "this tyranny which consists in a universal and uncontrollable desire to dominate everything regardless of order." This may explain why reductionism pushed to the extreme inspired some adherents to this form of intellectual imperialism to subordinate one discipline to another deemed in some way more noble, for instance biology to biochemistry, which ultimately leads to chemistry itself, if not physics. In this vein, we might mention the current trend in the sciences of the brain which claim to do a better job than psychology at explaining mental processes, or the efforts of some physicalists who want to assure us that biology could be rewritten on the basis of fundamental physics without having to introduce any truly new concept. Although in practice such ventures rarely succeed, the thinking behind them, based on a hierarchical ranking of various branches of knowledge, is pervasive and does enjoy some currency.

That said, if human thought is to have any meaning at all, it seems necessary to gamble that the universe can indeed be deciphered and to accept that our intellect has the means to fend off the insidious and ubiquitous hold of the multiple. The diversity of appearances does not necessarily prohibit unified concepts. That is what Pascal must have meant when he wrote in his *Pensées*: "Through space, the universe understands me and swallows me up like a point. Through thought, I understand it."

That the very idea of knowledge points to unity is a rather old concept. More than two thousand years ago, Xenophanes was already scoffing at the notion of a multiplicity of gods. Heraclitus proclaimed that "what is wise is one." Plato had proclaimed the unity of the universe in his *Timaeus*: "Our own view is that in all likelihood there is a single, divine world."[5] On the basis of that one observation, he worked out a detailed agenda with specific goals. They included the following:

1. To establish a geometry of the universe that would encompass within common concepts and figures the entire world and its elements (this geometry is described in Euclid's writings).
2. To consider that all transformations taking place in the perceptible world stem from structural changes in the arrangements of figures characterizing the elements: "We must proceed to inquire what are the four more perfect possible bodies which, though unlike one another, are sometimes capable of transforming into each other by dissolution."[6]
3. To state a universal principle of causality, according to which "everything that becomes or changes must do so owing to some cause, for nothing can come to be without a cause."[7]
4. To make a distinction between "two kinds of causes, one neces-

sary and the other divine,"[8] in other words, between the mechanical or errant cause and a cause subject to an intelligent design.[9]

5. To postulate that both types of causes intervened during the genesis of the world: "For this world came into being from a mixture and combination of necessity and intelligence. However, intelligence superseded necessity by persuading it for the most part to bring about the best result, and it was by this subordination of necessity to reasonable persuasion that the universe was originally constituted as it is."[10]
6. To consider that disorder, associated with errant causes, is an integral part of the fabric of the world.

Obviously, during the course of centuries, this agenda has lost most of its original features, but it has remained a constant source of inspiration and a steady guidepost. In the early part of the twentieth century, the school of logical positivism pledged in its 1929 manifesto the building of a "unitary science" that would take over the "sciences of the mind" and incorporate them under the umbrella of the "physicalism" of science, all the while affirming its indifference toward metaphysical speculations.[11] Whether this was incurable scientism (there can be no salvation outside science) or a concession to the metaphysical ideal of universality is unclear.

In any event, the dilemma of identity and diversity continues to cause even modern scientists many headaches, for reasons that are easy to understand. On the one hand, scientific theories are of necessity based on a stunning diversity of phenomena. On the other, the interpretive role of those theories can be fulfilled only through principles of unity and permanence. In a mass of specific observations, the scientist intuits the existence of a law; in the prodigious variety of phenomena, he senses a common thread, which he keeps tugging on until it breaks.

To unify is to make a reduction to identity. This quintessential act of theorizing is the hallmark of every formal scientific discipline. Starting from quite diverse data—with no apparent connection at first glance—the scientific process strives to extract a common theme and a single origin.

That is why the much-ballyhooed *grand unification*, which proposes to unify the four known fundamental forces in nature, is the avowed *telos* of physics, endlessly trumpeted with great hype in the media: The One often makes it to page one in the press. Yet the goal remains but a distant perspective. Two centuries ago, Immanuel Kant was convinced that Newton's mechanics could explain it all. Even a hundred years ago, Maxwell's theory continued to be taught on the basis of mechanical models.[12] At the present time, we simply lack the concepts and laws that would marshal in a single theory the universe considered as a whole in terms of its forces and ultimate constituents. In 1988, the British cosmologist Stephen W. Hawking concluded his book *A Brief History of Time* with the following comments, with quasi-mystical and supposedly prophetic accents:

> Even if there is only one possible unified theory, it is just a set of rules and equations.... However, if we do discover a complete theory, it should in time be understandable in broad principle by everyone, not just a few scientists. Then we shall all, philosophers, scientists, and just ordinary people, be able to take part in the discussion of the question of why it is that we and the universe exist. If we find the answer to that, it would be the ultimate triumph of human reason—for then we would know the mind of God.[13]

Is this the megalomaniacal delirium of an illuminated dreamer or sound predictions of imminent breakthroughs? Are we really on the verge of reducing the entire universe to a few equations? Such enthusiasm may sound suspect, but these comments and the message they convey do merit some attention. Are they merely an age-old incantation echoing an irrepressible metaphysical yearning, or do they herald something radically new? Has the outlook for success improved recently? We propose to offer some partial answers to these questions, and more importantly, to discuss their implications.

The present situation is marked by a peculiar and vexing paradox. On the one hand, science seems pervaded by relativism. An increasing number of philosophers have questioned the real contribution of science, accusing it of lacking a true methodology. They charge that aesthetic judgments, personal preferences, metaphysical prejudices, and subjective desires so taint the scientific method, at least in some of its phases, that it is no longer possible to vouch that it conforms to any rational process. Some even doubt the universal character of the rationality that drove science from the time of its Greek inception, and they go so far as to claim that it has no more legitimacy than ancient mythological beliefs.[14] At the other end of the spectrum, one also witnesses renewed efforts to develop all-encompassing theories, ironically in several different fields: A particular brand of biology claims to be zeroing in on the secrets of life; a specific type of mathematical physics professes to be deciphering the universe in its totality, including its birth and eventual fate.

Modern science keeps bouncing back and forth between these two extremes. A systematic and obsessive quest for the One coexists with an explosion of highly specialized studies. On one hand, the trend seems to be toward the Intellect with a capital I, and on the other, toward a proliferation of disciplines and subdisciplines. Even though the mind aspires to grasp the One, its methods of reasoning lead it to divide our body of experience into separate areas, almost without interconnections. What does the longing for unity really mean if the march of science drifts irresistibly toward what seems to give lie to unity?

In such an ambivalent situation, two distinct stumbling blocks stand in our way. If we claim to embrace the entire world, we risk criticism for being too dogmatic and for pursuing reductionism to excess. If, on the contrary, we retreat into specialized studies, we are likely to be taken to task for abdicating our responsibility. Between narrow-mindedness and pretension lies an almost insurmountable chasm. Astrophysics and more particularly cosmology are often blamed for using such double language. Do they amount to harmless musings, rather inconsequential because they are beyond verification, about a universe that is outside our grasp anyway? Or do they exemplify a relentless effort, soon to be completed, aimed at unveiling a grand scenario and confirmed by the most delicate measurements physics is capable of?

How far unity has progressed depends on where one looks. It has reached different stages in mathematics, philosophy, physics, biology, astronomy, and economics.[15] Our purpose here is to focus on the state of physics, for that is where the search for unity is currently being pursued the most actively and has been rewarded with the greatest achievements in the realms of both the infinitely large and the infinitesimally small.

Given the difficulty of such an undertaking, we will not dwell on the paradoxical fact—exposed for all to see on the very cover of this book—that it takes two authors to write about the One.

The
Quest for
Unity

The Greek Conceptions of Unity

By far the greatest gift is to be a master of metaphor. It is the only art that cannot be learned from someone else. It is also the mark of an original genius. For a true metaphor implies the intuitive perception of similarity in dissimilar things.
—Aristotle

Many of the Greek thinkers were inspired by monism. They aspired to explain phenomena on the basis of a single principle. Our senses may well suggest a world made of an infinite variety of things and phenomena, but for the purpose of trying to understand it, we have no choice but to introduce some semblance of order, in other words, some underlying unity. Natural elements, these philosophers believed, are not to be understood as isolated pieces disconnected from the rest. Instead, they must be amenable to integration into a broader explicative framework that gives them both meaning and functionality. Hence the conviction that, hidden deep within reality, there must exist some principle or fundamental substance from which all things can be derived. Out of this came a number of doctrines promoting one primordial element or another, the nature of which was not even necessarily inert or material.

The justification for starting this analysis by reviewing a few doctrines originating with the Greek thinkers is that they are the ones who planted in physics the seed of its eventual power. Yet, the Greeks do not have the corner on monism. For instance, the school of energeticism, promoted by Wilhelm Ostwald (1853–1932) and others, was a bona fide monistic metaphysics based on rejecting the concept of atom. The doctrine was rooted in a belief in the absolute primacy of energy as the agent governing the physical world in that it made it possible to understand, organize, and unify all observable phenomena. "Matter is an invention," wrote Ostwald, "and an imperfect one at that, which we forged to describe what is permanent in all vicissitudes. The effective reality, the only one that has an effect on us, is energy.... Nature as a whole appears to us in the form of spatially and temporally varying energy, distributed in space and time. We become aware of that energy only to the extent that it is transmitted to our bodies, and particularly to the sensory organs whose function is to detect it."[1] According to him, the triumphant expression of such a message was embodied in the principles of thermodynamics, especially the second law, which states that the entropy of a system can only increase.[2]

The Greeks looked at nature from a certain distance, although with profound admiration. Their intent was not to found a science in the mold of modern physics. The idea of dominating matter in order to transform it was alien to them. They perceived nature as the stuff of life, generating and confronting beings, evolving and changing endlessly, exploding into myriad forms, and unanswerable to logic and identification. These early thinkers were primarily concerned with observing and understanding the phenomenon of growth, ubiquitous, seemingly universal, not only in the vegetal and animal worlds, but in the realm of minerals as well, which in those days were believed to be subterreanean fruits growing in the belly of the earth and deposited on its surface through some continuous and unobtrusive process of percolation. Hence the etymology of the Greek word *phusis,* which is derived from the verb *phuein,* meaning "to grow."[3] The Greeks were simply looking for the primeval element of this natural growth, the fertile seed serving as the nutritive principle of life and of the entire cosmos. How was the world around us created, and how does it go on existing?

Water, the Primordial Stuff

Thales of Miletus (624–548 b.c.) held that water was the primordial element. "Water is the material cause of all things," he affirmed. It nourishes plants, quenches man's thirst, and is home to the fish that feed him. By falling from the sky and collecting at the bottom of wells, water irrigates fields and makes everything grow. Of all the known substances, water is the one that can take on the most varied forms: ice, snow, steam, cloud, liquid, not to mention rocks, which the Greeks believed to be made of frozen water (the word "crystal" comes from the Greek *krustallos,* which means "ice"). In river deltas, it appears as though it turns to earth. Elsewhere, it seems to spring forth from it. Water flows, disperses, soaks, transforms, infiltrates, fertilizes the soil, and carries ships. It is the liquid link that unites the whole with itself, saving it from being nothing more than a collection of things alien to one another. Besides, are air and fire not merely exhalations of water? It is universally recognized today that water is indeed an essential condition for life. Madison Avenue understands this fact quite well when it routinely associates water (preferably the bottled kind) with vitality.

Thales believed that the earth floats on water, having precipitated out on its surface much like silt precipitates in estuaries. With a little imagination, one can view this concept as the distant precursor of the theory of continental drift proposed at the beginning of our century by Alfred Wegener (1880-1930) who, upon noticing that the coasts of Africa and Brazil fit like pieces of a jigsaw puzzle, speculated that the two continents must have slowly drifted apart by floating on a liquid support. Great ideas never completely die out: Water is an idea that refuses to fade away. Nietzsche acknowledged in his *Philosophy in the Tragic Age of the Greeks* the momentous impact of the "gigantic generalization" carried out by the greatest philosopher to come out of Ionia: "Thales saw the unity of Being, and when he set out to describe it, he spoke of water."[4]

Indeed, for the first time ever, a single and all-encompassing explanation had been put forth to account for everything. This economical approach was a departure from a long-standing tradition that relied on a multiplicity of arbitrary causes and ad hoc explanations. This in no way means that prior explanations were all incoherent. Myths had their own logic and fit together with a certain plausibility. In his *Philosophy of Symbolic Forms,* Ernst Cassirer showed how even a universe full of myths is invariably characterized by a chain of interlocking steps. Such a chain may be "independent of the laws of empirical thought," but it conforms to laws nonetheless and reveals "a kind of original and autonomous structure," that is to say, a "theory," provided that the meaning of the word is generalized.[5] For instance, Hesiod recounted in his *Theogony* how an initially chaotic world was gradually unified through the sovereign power of Zeus: "At the very beginning Chaos and the Gaping Abyss were born, but next came Gaia the Earth with her broad bosom. . . ." It all seems to mesh together seamlessly, even with a certain elegance. But by invoking so many different causes and divinities, myths ended up creating a narrative more than an explanation. They provided a description rather than a system. The interconnection, orderly as it may be, of too many causes ends up looking like randomness.[6]

Thales expressed for the very first time three ideas that were to prove seminal. First, he raised the question of the material cause of all things. Second, he insisted that the answer not rely on mythology and folklore or on gods that were evidently far too anthropomorphic.[7] Lastly, he postulated that in the end it must be possible to reduce everything to a fundamental substance of a more profound nature than any other. Thales and his followers legitimized the will to frame what amounted to scattered bits of knowledge into a rational and coherent system freed from bondage to the occult. Myths and fables were gradually discredited and came to be replaced by the reasoning of philosophers. Neither the multiple creations of Gaia nor the authoritarian interventions of Zeus could continue to be satisfactory explanations. With his primordial water, Thales had ushered in the era of the rational mind.

Air, Fire, and Earth

In the sixth century B.C., Anaximenes (ca. 550–480 B.C.) propounded yet another cosmogony founded on the idea that the principle of the universe is air. Before exhaling our last breath, do we not indeed spend our entire lives drawing in this life-sustaining stuff that is all around us, according to a rhythm that orchestrates the interplay between the external world and ourselves? According to Anaximenes, the whole world is suffused with a vital flow that brings life to its various forms and coordinates its many parts. Our soul is a form of air that sustains us. The Earth, thin and delicate, is carried like an autumn leaf by the currents of an invisible atmosphere. The wind, created by circulating air, fills the sails of ships. By condensing, it forms clouds that can themselves condense further to bring rain and even create stones. By rarefying, air turns to fire. And what about snow, one might ask? "It is a breath of air imprisoned in humidity." At the cost of a few contortions, monism can come up with an answer for everything.

Anaximenes's greatest merit is to have shown that once a particular primordial element is chosen, it is possible to deduce all other known components of the world. It was not all that easy, even with all sorts of metaphors, to reduce all known phenomena to arguments involving nothing more than air. In Anaximenes's defense, it took no fewer than twenty-four centuries to finally establish that air—as well as water—is actually a composite mixture, which disqualifies it from the title of primordial element.

At about the same time, Heraclitus of Ephesus (ca. 550–480 B.C.) picked fire as the element explaining all phenomena in the universe: "All things are exchanged against fire, and fire against all things." Fire lights the world and rules it. By condensing, it generates water; and by condensing even further, it becomes earth. Fire is the principle of heat and life; it is the flame that feeds every passion, and which all lovers pledge to each other. For Heraclitus, it was all quite simple: "Lightning governs the universe," which itself is nothing but an eternal fire that will never be completely extinguished. Destined to a worldwide blaze and final conflagration, it will consume itself, not to disappear forever, but to rediscover in this ultimate inferno the principle that will enable it to be reborn from its own ashes.

The world, conceived of as a cyclical brazier, would evolve according to a recurring pattern, fire serving as the unifying link between successive phases. Some will undoubtedly detect in such a picture the seed of the doctrine of eternal rebirth, which would prove so appealing to the Stoics, and which Nietzsche would later describe as the only thought with a chance of defeating nihilism.[8] Others might even make a parallel between this concept and a particular version of the big bang model, which holds that the universe is subject to a series of expansions and contractions, in which the transition between successive cycles is characterized by enormously high densities and temperatures.

The concept of becoming occupies a central place in Heraclitus's philosophy, to the point of implying a categorical rejection of the notion of being, so dear to Parmenides. Fire becomes the paradigm of things that evolve. No form of permanence can ever fix it. If fire fascinates us so, it is because not a single one of its flames can ever be captured. "I cannot imagine anything more fleeting and alive than fire...," wrote Jean de la Fontaine (1621–1695) in one of his celebrated fables.[9]

Questions of unity invariably bring up the issue of completedness. How is it possible to reduce every possible phenomenon to a matter of mere flames? To resolve this dilemma—confronted by any monistic philosophy—Heraclitus came up with an ingenious idea. He asserted that a "struggle between opposites" presides over nature. The "electricity" between antipodes supposedly constitutes the true unity of the world, which would be at once one and multiple. As the driving force of becoming, it is to be viewed as principle and law. "Sea water, to take one example, is both very pure and very impure; it is a nutrient for fish; but for men, it is undrinkable and harmful."[10] Nature, asserted Heraclitus, manages to sublimate antagonism into harmony, much like music created by the lyre is rooted in the tension between strings and wood. "Opposites are brought into consonance and a beautiful harmony is engendered by what is contrary." Heraclitus repeatedly emphasized that all things originate in conflict and amount to accommodations between contrasts, precarious equilibria between opposing forces, manifesting themselves in endless change. "We must realize that discord is common to all things, that justice results from struggle and that all things are born and disappear through strife."

One could argue that many aspects of modern physics indeed appear to conform to Heraclitus's doctrine. It often seems that mutual opposites are fated to be the source of conceptual developments, as if no progress were conceivable without an infusion of dichotomy. The concept of quantum complementarity, for instance (to which we will return later), is directly inspired by such a notion. It asserts that waves and particles are two descriptions—logically incompatible, but inextricably intertwined just the same—of quantum reality. Pushing the metaphor a bit further, other examples of interplay between opposites can be found in the world in the infinitesimally small as well. A photon, for instance, can be created by the annihilation of an electron and a positron, and in the reverse process, a photon can disappear to generate an electron-positron pair.[11] The symmetry between matter and antimatter could also be construed as reflecting the unity of opposites. More generally, any discourse exalting the virtues of paradoxes is never too far removed from Heraclitus's philosophy.

Yet, the living world beats physics by miles when it comes to demonstrating the power of opposites. In one of his books, Arthur Schopenhauer describes the strange case of an unusual animal called the Australian ant-bulldog: "When you cut it in half, a battle ensues between head and tail; the head clasps the tail, which frantically uses its sting to try to evade the head's bite. The fight can last half an hour, until death follows, unless other ants carry away the two pieces."[12] Inasmuch as the story is true, the ant-bulldog, belligerent to its very core, illustrates the ferocious traits the dialectic of the One and the multiple can assume when it leaves the province of pure philosophy.

Werner Heisenberg (1901–1976), one of the founding fathers of quantum physics, observed in *Physics and Philosophy* that in some regards modern physics has much in common with Heraclitus's doctrine: "If we replace the word 'fire' by the word 'energy,' we can almost repeat his statements word for word from our modern point of view. Energy is in fact the substance from which all elementary particles, all atoms and therefore all things are made, and energy is that which moves.... Energy can be changed into motion, into heat, into light, and into tension. Energy may be called the fundamental cause for all change in the world."[13]

Plato, on the other hand, took a dim view of those who, like Heraclitus, engaged in game playing with opposites, with no apparent justification. Such a proclivity was, in his opinion, the mark of superficial thinkers and novices in the art of argumentation. The combination of antipodes tells us nothing, he argued; it forms a net so loose as to let virtually everything through. That is why Plato consistently and relentlessly ridiculed anyone who claimed that the One is multiple, and the multiple One, or that the Being is the non-being, and vice versa. "But to show that 'the same' is other in some way no matter what, and 'the other' is the same, and the big small, and the similar dissimilar, and in this way to take pleasure in always putting forward the contraries in one's speeches, this is not a simply true examination, and it shows as well that it is the fresh offspring of someone who just now is getting his hands on the things which are."[14] That did not prevent Plato from considering dialectics the most important part of any philosophical endeavor!

In summary, water, air, and fire were in turn viewed by the early Greek thinkers as playing a primordial role in the organization and evolution of the world. Empedocles of Agrigentum (490–435 B.C.) brought some modicum of agreement among these various doctrines before he threw himself, according to legend, into the crater of Mount Etna to prove his reputation as a god.[15] He maintained that all of our earthly world could be explained on the basis of four elements of equal importance—water, air, fire, and earth (which Xenophanes had in the meantime proposed as the primordial stuff)—and two principles that united or dissociated them—love and hatred. Introducing several primordial materials, no matter how few, offered much greater flexibility because of the possibility of their various combinations.

The preeminence of these four elements was to permeate the cosmologies of the Middle Ages and the Renaissance. For example, the affinities between natural elements and corporeal humors made it natural to regard man as part of the world and, at the same time, the world as a human organism. Moreover, it is not too far-fetched to consider this quartet of primeval materials, made up of elements endowed with qualities that enable them to interact, as presaging our modern views on the structure of matter. Admittedly, the fundamental constituents recognized by modern science bear little resemblance to water, air, earth, or fire. Instead, we now have six quarks and six leptons, grouped in three families with similar structures. And, perhaps sadly for the poetry of things, it is no longer love and hatred that cause them to interact, but four fundamental forces.

The Concept of Fundamental Substance

Early on, the Athenian philosopher Anaxagoras (610–547 B.C.) had proposed the intellect as principle and organizing force of the universe. Conceived of as a subtle substance, it would render all things liable to a mechanistic explanation. Anaxagoras distinguished two kinds of beings. First, the elements proper, which are actually mixtures because "in everything is a part of all other things," and, second, our reasoning (the so-called *nous*), which is the architect of the world. *Nous* stands by itself, pure and unadulterated. Faced with the chaos of the elements, it restored order to it. With his doctrine of the organizing *nous*, Anaxagoras was to exert a decisive influence, notably on Socrates.[16]

During the same period, Anaximander (610–546 B.C.), a pupil of Thales's, rejected water, or any other known substance for that matter, as primordial stuff. He could not accept that fire could come from water, when every ordinary observation shows these two elements to be rather incompatible. Since no element is privileged, it is impossible to consider any one of them more suitable than any other as substratum. In order to get around this difficulty, Anaximander taught that the primordial stuff was of a different essence. It had to be infinite, eternal, and ageless, so that it may embrace the world in its entirety. Only something far richer and more all-encompassing than anything else could possibly guarantee the equilibrium and harmony of the universe. In the end, this primordial substance was just an abstraction, the nature of which was more ethereal than any material substance, and the existence of which carried within itself the potentiality for the existence of all the others. Anaximander gave it the name *apeiron*, which means "boundless." He described it in terms of adjectives traditionally reserved for god-like entities, such as uncreated, immortal, all-encompassing, and so forth. It seems that the One, with a capital O, often tends to bear an uncanny resemblance to God.

That the world can be structured on an abstract principle is an idea that science would take note of and would ultimately use to great benefit. It constitutes a premonition of the fruitful role to be played by the notion of the infinite in developing the science of mathematics. It also heralds the concept of a universe encircling totality, the introduction of which, in the seventeenth century, would essentially launch physics on its modern course and ensure from the outset its unifying potential.

Anaximander explained how the *apeiron* contains in its midst the entire concrete future of things. It is the effective cause of all that is born, matures, and dies off, the ultimate origin of all individuals, who may segregate themselves from it, but to which they all end up returning at the time of their final dissolution. The *apeiron* transforms itself to give us all the substances we are familiar with. In the only extant portion of his work entitled *On Nature*, Anaximander wrote: "The apeiron is the principle of all things that exist.... That from which all things derive their existence is also what they return to upon their destruction, in accordance with the order of necessity. Things mutually justify their own existence and correct their injustices according to the flow of time."[17] Evidently, Anaximander thought of existence as something of a loss, a renunciation of a primitive source. Nietzsche, perhaps Anaximander's most enthusiastic exegete, would later declare that "all becoming is an emancipation guilty of forsaking the eternal Being, an iniquity that must be atoned for with death." Upon their birth, things cast themselves off the original unity in order to fulfill their own destiny. But in doing so, they commit an impious act for which, in the interest of justice, the ultimate punishment must be exacted. To be born becomes tantamount to dying.

For Anaximander, the struggle never ends between warm and cold, fire and water, moist and dry, as they take turns predominating. One form disintegrates into another but never remains static for long. Any victory of one over another is but transitory. All the parts keep changing, while only the whole remains immutable (centuries later, by establishing that energy is globally conserved even though it can assume multiple forms, physics will adopt a very similar view). Anaximander thus places being and becoming on opposite sides of the equation. The primordial stuff, infinite and ageless, forms an undifferentiated being that degenerates into various forms destined to struggle endlessly with one another. Multiplicity results from falling into individuality, and the very process of becoming is to be considered a degradation of being.

The Negation of the Multiple and of Motion

The problem of the connections between the One and the multiple brings us inescapably to Parmenides (544–460 B.C.). He declared quite simply: "Being is, non-being is not."[18] Such a statement implicitly rejects the multiple and becoming, both of which Parmenides equated with non-being. Since they did not exist, any speculation about them could only lead to contradictions, hence to errors: "What does not exist cannot be known or even described; for what can be thought of and what can exist are one and the same."[19] Only the One, necessary, eternal, and unengendered, exists, and there is no such thing as becoming or disappearance. By the same logic, Parmenides also denied the existence of void, which he assimilated with non-being. He further rejected change and even movement as pure illusions, which would earn him the wrath of Aristotle. In Parmenides' mind, any change contravenes the innate propensity of the mind to accept identity and permanence.

The lesson was not lost on physics, even in its most modern form. Does the perenniality of its laws, supposedly immutable and invariant, not imply the eradication of time? The moment we deal with processes that evidently have a past history and have evolved, it is only to try to discern substance and forms, rules and laws that are themselves independent of time. The very goal of physics is to understand what changes in terms of what is permanent. It accomplishes this by establishing laws that have been set free of the dictates of time, even though these same laws may well describe time-dependent phenomena.

Admittedly, physics does not really have much of a choice. How could anyone ever construct a theory starting from fleeting concepts? What would physical laws look like if they were based on notions applicable one moment and not the next? How could unity be preserved if concepts were constantly changing? Science rests implicitly on the postulate that the relations implied by laws are steady and constant. If so, what, if anything, can we know about how laws might evolve with time? The question happens to be a crucial one in cosmology, which deals with a universe in expansion. It seems reasonable to follow Henri Poincaré's lead in decreeing that "laws not be functions of time, even if facts were to subsequently prove us wrong and force us to broaden these same laws, and possibly even amend them."[20] From a pragmatic standpoint, though, the challenge is to find the most economical strategy, which is probably to assume that what changes with time is not the laws themselves but the "universal" constants that enter into them.[21]

Atomism

Atomism developed in parallel with monistic doctrines. The theory was born on the shores of the Mediterranean. It aimed to answer certain questions about the beginning and end of the cosmos, about the unity and diversity of material objects, and about permanence and change. Gaston Bachelard (1884-1962) would later call it a "doctrine of diminutive objects," or a "metaphysics of dust."[22] The theory was originally developed by Leucippus (fifth century B.C.), a contemporary of Zeno of Elea, and later amplified by Democritus of Abdera (ca. 460–370 B.C.), a contemporary of Socrates. It explains the world by resorting to atoms, entities without sense qualities, which the intellect alone is capable of comprehending, and to void. These tiny units of matter are, in a sense, being's last refuge from reduction to its primordial constituents. They are undissociable (*a-tomos*), eternal, filled, solid, and unlimited in numbers.

In this picture, the qualities we might today refer to as secondary (such as color, odor, flavor) have but a subjective existence. Only the geometrical properties of atoms (size, shape, position) are real. When in rapid motion, atoms undergo mutual collisions that result in the formation of aggregates distinguishable by the shape, order, and position of the constitutive atoms, rather like words are composed of letters. According to Democritus, who wanted "to feel in the world as if he were in a brightly lit room" (to use Nietzsche's phrase), the entire world—including the heavens, the earth, animals, and even the soul—was made of these aggregates and nothing else. Final causes were specifically excluded. In the process, Democritus relegated the gods to the cloakroom, since no intervention on their part was required to explain the universe. The fear they once inspired in us being groundless, all our superstitions could be safely discarded.

It is no doubt unnecessary to underscore the success enjoyed by this doctrine, at least in its modern incarnation. Yet, it was not until the early part of the twentieth century that the atom itself became the object of direct studies in physics.[23] Its entrance onto the stage may have been long delayed (it remained dormant for twenty-five centuries!), but it came in with a bang. It triggered nothing less than the quantum revolution. It quickly became apparent that the atom was in fact quite a complex object, devoid of almost every incipient quality Democritus and his followers had ascribed to it. Between antiquity, when it was merely an abstract idea, and the modern era, when it became an object of scientific study, the atom has inspired extraordinary intellectual jousts spanning two millennia.[24] The enduring attention it has commanded among the best minds seems to vindicate Bergson's view that human intelligence, "shackled as it is by geometry," cannot break free of the issue of the divisibility of matter.

Indeed, most great thinkers—be they theologians, philosophers, or scientists—have taken an active part in this great debate, as if atomism had been a powerful magnet for human thought: Epicurus (341–270 B.C.) invented the concept of *clinamen*, "the tiny swerving of the atoms at no fixed place and no fixed point of time"[25]; Aristotle rejected atomism outright on the grounds that void itself could not exist; Descartes did likewise; Gassendi resurrected the idea of elementary entity; Giordano Bruno added to it spiritualistic undertones; Leibniz initially believed in atoms and void because "that is the most satisfactory picture for our imagination," but he later realized that it is impossible "to find the principles of a genuine unity solely in matter or in what remains inert, since under these conditions everything is but an infinite collection of separate parts"[26]; Hegel too had a negative opinion of corpuscular theories; Schopenhauer condemned the reductionism they imposed. Not to mention Kant, Nietzsche, Maxwell, Marx, Comte, and many others. We cannot list them all.[27]

The theory of atoms would ultimately lead to the great unifications of twentieth-century physics.

The Plurality of the One

Greek philosophy gave rise to a host of metaphysical entities, both visible and invisible. In spite of their individual characteristics, these entities had at least one property in common: Each one of them, constructed and promoted for its unique power of synthesis and for its ontological merit, purported to account all by itself for the great diversity of the perceptible world. The atom, to take a specific example, was touted as the substance underlying all phenomena, causing effects that are observable all around us, and the real permanence hidden in changing appearances. It is a principle truly able to yield a synthesis of the physical world, and much more. All the way to its most recent developments, physics would repeatedly draw from this reservoir of images and concepts.

The pre-Socratic philosophers had foreseen the plurality of the concept of unity. Accordingly, they envisioned the One either as a principle rooted in numbers or as a principle generating and organizing the world. The systems they proposed offered a vast array of choices, including pluralism and mobility (Heraclitus), pluralism and divisibility (Democritus), unity without diversity (Parmenides and the Eleatics), and even diversity without unity (Cratylus, a disciple of Heraclitus, believed that the flux of existing things is so overwhelming that reason is powerless to apprehend it).

Yet, for all the achievements of the Greek philosophers intent on constructing an overall description of the world, it is easy to forget the criticisms that such proposals immediately provoked. Plato, for instance, wrote about the deep sense of disillusionment his teacher Socrates experienced upon reading Anaxagoras. Socrates lamented:

> I was glad to think that I had found in Anaxagoras a teacher about the cause of things after my own heart, and that he would tell me, first, whether the earth is flat or round, and then would explain why it is so of necessity, saying which is better, and that it was better to be so. If he said it was in the middle of the universe, he would go on to show that it was better for it to be in the middle.... This wonderful hope was dashed as I went on reading and saw that the man made no use of Mind, nor gave it any responsibility for the management of things, but mentioned as causes air and ether and water and many other strange things.[28]

Socrates came to the conclusion that scholars in the philosophy of nature (in a broad sense, this means physics) often pretended to know "the cause of all things" when in fact they were a long way from having all the answers. "Imagine not being able to distinguish the real cause from that without which the cause would not be able to act as a cause. It is what the majority appear to do, like people groping in the dark; they call it a cause, thus giving it a name that does not belong to it. That is why one man surrounds the earth with a vortex to make the heavens keep it in place; another makes the air support it like a wide lid."[29] Aside from the Sophists, Socrates was the first to sound a warning against the pretensions of the human mind and to bring back down to earth philosophers blinded by premature speculations.

Blaise Pascal, too, was convinced that human knowledge is incapable of comprehending the first principles, let alone the unity that supposedly enfolds them. He shared Socrates' wholesome caution and skepticism: "Strangely enough, they [men] wanted to know the principles of things and go on from there to know everything, inspired by a presumption as infinite as their object. For there can be no doubt that such a plan could not be conceived without infinite presumption or a capacity as infinite as that of nature."[30] The comment might serve as a healthy reminder to those who even today claim they can explain, with a couple of ideas, what the world is made of, where it comes from, where it is going, how it works, and why it is the way it is!

Unity through Numbers

Out of the multiple, ideas spring forth.
—*Henri Michaux,* Affrontements

The reflections of the Pythagoreans on numbers around the fifth century B.C. also center on the issue of unity. Starting with the concept of the One, they sketched the outline of a coherent model of the world. If things are assimilated with numbers, or so they claimed, then everything can be derived from the One, and the very affirmation of the multiple becomes a basis for the notion of unity. Since each number amounts to unity multiplied, the process of multiplication itself can in principle encompass all numbers as well as all things. The multiple lives hidden and folded into the One. Moreover, if numbers are the reason for everything, it is but a short extra step to attribute to them a moral value. And so, justice came to be represented by the numbers 4 or 9, which are squares and evoke perfect harmony.

From the Pythagoreans' point of view, it is quite simply the genesis of numbers that reveals the laws of the universe and of human thought, not to mention those of music or the structure of constellations. "Everything there is to know has a number. Without numbers, we cannot understand or know anything," declared Philolaus in the fifth century B.C. Numbers, invested with considerable symbolic power, are the repositories of unity for two reasons. First, because they constitute a utilitarian and operational concept with which everything else can be thought of through extension and multiplication. And second, because their very nature suggests that they can be assimilated with the fundamental unit, in other words, with the One in the arithmetical sense.

The Pythagoreans' concepts have not fallen into oblivion. The idea that numbers have a real existence of their own, with nature simply reflecting indirectly their properties, has enjoyed an enduring popularity. The fascination elicited by Pythagoreanism had fertile repercussions on the development of arithmetic. It gave credence to the idea that truth is inherently connected with harmony, and it inspired a geometrical vision of the world based on arithmetical principles. It continued to evolve under various forms right down to the time of Galileo, Newton, and even Maxwell, Einstein, and Dirac. In fact, both Einstein and Dirac felt that the aesthetic mathematical appeal of a physical theory was not just to please the mind. It was also an indication—indeed perhaps the best there is—of its validity.

Mathematical reasoning has the ability to venture beyond facts. In a 1952 letter to Maurice Solovine, Einstein clarified the distinction he made between "constructive theories" and "theories of principle." The former seek to "uncover true laws through an effort of elaboration" (*durch...konstruktive Bemühungen*); they derive their value solely from empirical confirmation. The latter aim to "discover some general formal principle"; they derive their value from experience, obviously, but most of all from their "internal perfection" (*innere Vollkommenheit*).[31] By virtue of their intrinsic properties, they enjoy empirical validity as well.[32]

In his Spencer Lecture, delivered at Oxford in 1933, Einstein stressed the importance to be accorded to formal beauty: "Experience can of course guide us in our choice of serviceable mathematical concepts; it cannot possibly be the source from which they are derived; experience of course remains the sole criterion of the serviceability of a mathematical construction for physics, but the truly creative principle resides in mathematics. In a certain sense, therefore, I hold it to be true that pure thought is competent to comprehend the real, as the ancients dreamed."[33]

Born in antiquity, the concept that the universe is founded on numbers was thriving in the Middle Ages, as Bernard Ribémont pointed out:[34] "Numbers are given a symbolic dimension and endowed with original properties participating in the creation of the universe and matter." So much so that "through practice and the study of numbers and their combinations, man is called upon to try to rediscover the properties of the world and nature."

Christianity had long embraced such concepts and claimed them as its own. For instance, God's mastery of things mathematical is extolled in the Old Testament: "You [the creator] have arranged all things by measure and numbers and weight."[35]

The twelfth century even witnessed the advent of a "Christian arithmology," based explicitly on the reality of numbers and oriented toward the reading of sacred texts. It differed from traditional arithmetic in that, instead of establishing formulas, it sought to extract from numbers as much information as it possibly could in terms of symbolic correspondences. Its position was that "the more substance is extracted from numbers, the better the chances to understand the significance of those numbers, even if outside the domain of mathematics," as Bernard Ribémont reminds us. For instance, the number 3, being a prime number and therefore "undissociable," corresponded to a perfection strongly evocative of a divinity. It came to represent the mystery of the divine trinity, itself associated with the heavens. The number 2, on the other hand, obviously less perfect since it is divisible by two, symbolized the duality of soul and body and stood for the flaws besetting the earth.

As can easily be appreciated from this simple example, such an approach has very little to do with mathematics, except for borrowing a few selected members of the infinite sequence of integers. Proceeding by analogies, correspondences, and other potentially risky associations, by mixing symbolism, arithmetic, astronomy, sacred texts, and natural laws, this type of doctrine hardly qualifies as a rigorous science. More aesthetic than effectual, it fell in disfavor after the genuine science of algebra, which reenergized the tradition of the early Greek mathematicians (such as Aristotle, Euclides, and Archimedes), began to make inroads in the twelfth century of our era, notably with the work of the Arab mathematician al-Khwarizmi. Neither the symbolism of numbers nor the philosophical and religious implications of their presumed "power" were able to compete with an authentic mathematics based on systematic numerical manipulations. The mysticism of numbers, even though it managed to introduce a limited amount of order in some concepts, provided not much more than a vague and arbitrary form of numerology.

The Harmony of the World and the Poetry of Order

While it is undeniable that mystical approaches had virtually no direct scientific impact, the quest for harmony in nature has nevertheless been a constant driving force behind physics research. Harmony is "the poetry of order," as one of Balzac's characters put it in a letter to the Dutchess of Langeais. It is probable that science would have made very little progress had it not constantly been goaded on by aesthetic or metaphysical considerations. Whether it takes the form of a conscious and deliberate search for harmony, of the type advocated by Kepler, or a kind of "wakeful dream" similar to what Einstein often talked about, such motives do play a crucial role—certainly ambiguous at times, but indisputable nevertheless—in the process of scientific discovery. When it invents itself, science finds its source outside its immediate sphere. As such, it has become dogmatic and parochial to insist that metaphysics had nothing to contribute to science.

It is probably Johannes Kepler (1571–1630) who best epitomizes the transition between mystical exploration of the harmony of the world and genuine scientific inquiry. Some of his successors, notably Galileo and Newton, are often hailed as the founders of modern science. But their own achievements depended critically on Kepler's genius. Some critics may lament the fact that Galileo chose to malign what he considered Kepler's undue mysticism, rather than do the honorable thing and pay tribute to his illustrious predecessor.

Kepler dedicated himself to uncovering the hidden harmony of the world, which he viewed as a reflection of God's perfection. He believed that this perfection must manifest itself in different forms, more or less metaphorical, and changeable so as to accommodate the evolution of thought. One such form was the traditional complement of spheres and circular orbits. Kepler used geometry in a very original way by inscribing the five perfect solids favored by the Platonists between the orbits of the planets in order to account for their relative sizes.[36] Such solids were the best candidates in the entire geometrical know-how of the time to embody the notion of symmetry. As such, they seemed a most promising avenue toward harmony.

But Kepler himself realized that this geometrical analogy was inadequate. Indeed, it lost all relevance after he discovered that planetary orbits were actually elliptical. Far from being discouraged, Kepler then went on to attempt to express harmony through music, rather than through numerical ratios or geometrical figures. The idea of a world in tune with musical chords was not new. The Pythagoreans, Plato, Aristotle, and many others had already made references to a universal harmony created by celestial spheres. As early as 55 B.C., Cicero had explained in *Scipio's Dream* why we fail to hear the corresponding sounds: Our ears are deafened by the din of the universe, much like the people living on the shores of the river Nile have become oblivious to the roar of its falls.

According to Kepler, the faster a celestial body moves, the higher the pitch of the sound it emits. Therefore, a planet emits notes determined by the characteristics of its elliptical orbit. That principle makes it possible to associate with each planet specific musical ratios and, ultimately, particular notes.[37]

In any event, harmony reigns supreme throughout the heavens, and that is a direct result of an explicit divine will. This harmony happens to be complex, owing to the elliptical nature of orbits. But that did not worry Kepler at all—quite the contrary. He actually deemed it superior to the blander harmony that would prevail if orbits were perfectly circular, as had been believed before Kepler's time. Such a richer arrangement leads to a "harmonic beauty [that] surpasses the beauty of simple geometry."

In his work *The Harmony of the World*, Kepler attempted to uncover the secret of the universe through an ambitious synthesis of geometry, music, astrology, and astronomy. Everything converges, he maintained, to exalt a consonant world. The universe is supposed to sing with exquisite perfection, more beautifully even than a choir.

His reflections blazed the trail to be followed later by Galileo, Descartes, and Newton in their own mathematical constructs. They would expand on Kepler's incipient theory of harmony and eliminate a few of its arbitrary elements. As it turns out, Kepler's ideas were not as naive as they may appear. Although we still have no complete understanding today of what determined the arrangement of planetary orbits (see the discussion of the Titus-Bode law below), phenomena of resonance between planets may be involved and are being explored. The term "resonance" is obviously not to be taken in its musical sense. Nevertheless, it does imply, through the bias of dynamics, specific relations between orbits that are redolent of tonal resonances. Even though it makes no explicit reference to musical concepts, modern physics does apply a methodology not unlike that used by Kepler.

Kepler carefully laid the groundwork for subsequent efforts at geometrization. His first law, quite bold for its time, discarded circular planetary orbits (by then these had acquired the force of dogma because of their supposed perfection) in favor of elliptical trajectories, which had been studied long before by geometers in ancient Greece and Alexandria. It paved the way for the unification of all motions to be achieved later on by Galileo and his successors. His second law is the celebrated "law of areas": It stipulates that as a planet proceeds along its orbit, the area swept out by its motion is proportional to time. This was a radically new description of the motion of planets. In particular, it implied that the farther away from the sun, the lower the speed, in distinct contradiction to what had been the common wisdom up to that point. In fact, these are clearly mathematical laws, and Kepler himself claimed as much: "My aim is to show that the heavenly machine is not a kind of divine, live being, but a kind of clockwork,...insofar as nearly all the manifold motions are caused by a most simple, magnetic, and material force, just as all motions of the clock are caused by a simple weight. And I also show how these physical causes are to be given numerical and geometrical expression." The sentence is quoted by Arthur Koestler, who emphasized the portentous impact of this rudimentary attempt at mathematization: "He [Kepler] had defined the essence of the scientific revolution."[38]

Kepler used the entire arsenal at his command to express, however tentatively and intuitively, the harmony he so believed in. He deserves enormous credit for having been able to formulate with such precision the laws that bear his name. One must appreciate that in those days the mathematical tools—those that describe what is now called symmetry theory—that would have made his task so much easier did not yet exist. As soon as they became available, other physicists started using them to express more rigorously the Keplerian notion of harmony. Science appropriated that part of metaphysics that had contributed to its own birth and, resorting heavily to notions of symmetry duly couched in a mathematical language, recast in a far colder formalism the concept of harmony, which in the process lost much of its idealistic character.

Whether the drive toward harmony was out in the open, as with Kepler, or more subdued and subtle, as with many of his successors, it always remained very much alive. Its unifying power is unquestionable, as evidenced by a number of pivotal breakthroughs it inspired in the evolution of physics.

The mathematician Pierre Cartier suggests that the paradigm of the quest for harmony is the Titus-Bode law.[39] It is embodied in an empirical formula, proposed in the latter part of the eighteenth century, that expresses the property that the radii of the orbits described by successive planets form an approximately constant ratio, roughly equal to two. Astronomers have long felt that it must reflect a more fundamental law yet to be discovered. They have endlessly speculated whether it implies an underlying harmony and, if so, what its nature is and where it is hidden. Many different explanations have been proposed to justify this intriguing empirical relation. Some are based on the notion of resonance; others involve turbulence or scale invariance. A few even resort to the so-called golden number, the ultimate symbol of harmony if there ever was one. Any one of these concepts had the potential to become a unified description of an entire class of phenomena. Yet, in the case of the Titus-Bode law, none proved compelling enough, even though the law has since been shown to be even more general than initially believed. (Similar laws have been shown to apply to planetary satellites as well.) Had one of them emerged as a winner, its underlying ideas would instantly have acquired an irresistible unifying power and would have become the basis of a new synthesizing mathematical description. The hope never fades that an empirical law initially defying explanation will eventually make sense in the framework of a broader and more general theory.

Cartier cites another famous case, involving the frequencies of a subset of the spectral lines of hydrogen, known today as the Balmer series. Although the pattern followed by these frequencies started out just as puzzling as in the case of the Titus-Bode law, the mystery ended up being resolved so successfully that it literally revolutionized physics in the early part of this century. The Swiss Johan Balmer (1825–1898), a man highly adept at manipulating numbers and probably inspired by his own Pythagorean convictions, showed in 1885 that the frequencies ν of these lines, which form an infinite series, could be described by a single empirical formula written as $\nu = R\,(1/2^2 - 1/n^2)$ where n is an integer equal to 3, 4, 5, etc., and R is a constant. This very simple equation constitutes a flash of genius because it predicted with considerable accuracy an infinite number of frequencies; yet it came to light at a time when only four of those were known, and only empirically at that. Not long thereafter, in 1890, this empirical formula was generalized by the Swede Johannes Rydberg, who proposed including all terms of the form $\nu = R\,(1/m^2 - 1/n^2)$, where both m and n are now series of integers. At first, these formulas were very much like the Titus-Bode law, in that the involvement of integers strongly suggested the likelihood that hidden within them was a fundamental harmony to be discovered. The riddle was finally solved by Niels Bohr, who proposed in 1913 a model of the atom that accounted perfectly well for these tantalizing empirical rules. He even managed to calculate the constant R, known as the Rydberg constant, in terms of other fundamental physical constants.[40] In this particular case, the urge to find a theoretical explanation for empirical laws was rewarded beyond any hope.

These examples suggest a possible epistemological model for how ideas evolve in physics. It all starts with an empirical law (e.g., Kepler's, Titus-Bode's, Balmer's, or Planck's) describing a particular category of phenomena. This first step has a unifying character from the outset, since it offers a common description of situations that may well be related but nevertheless appear quite distinct at first sight. The law so established is all the more credible since it typically leaves a number of empty slots in the array of phenomena, slots that in time get filled.

The crucial question is then to determine if the law is the expression of something deeper than a mere phenomenological description. Mathematicians can always fit a number n of data points to a sufficiently complex function (for instance, with an nth-degree polynomial). This establishes a "local" empirical law. But such a result is completely trivial. Until further notice, that effectively remains the status of the Titus-Bode law. On the other hand, it may harbor a much more profound significance, with almost magical implications, heralding a full-fledged theory that will give everything meaning. Things get unified and the law begins to make sense in a much broader context. Empirical at first, it ultimately comes to be perceived as reflecting a deeper harmony in the new theory. That is precisely what happened with Rydberg's law of energy levels in atomic physics. The same can be said of the blackbody law established by Planck in 1900, whose heuristic power was such that it almost singlehandedly toppled the principles of classical physics.

The historical impact of a simple formula can be momentous. What is it that guides the one who makes the discovery? Is it chance? Intuition? Or is it a deep-seated conviction that some harmony must lie hidden behind any hint of orderly pattern? The jury is still out on this question.

Particle physics provides a particularly striking example of a phenomenological view based on harmony, expressed as it is in terms of symmetries. Recent developments in the quantum theory of fields exploit analogies between the behavior of particles (or, more precisely, the structure of the interactions governing their behavior) and mathematical symmetry groups.

Whereas identifying and defining the various fundamental particles started out by focusing on the corpuscular nature of these objects, the classification scheme suggested by modern theories amounts to a search for harmony, since their objective is to discover applicable symmetries. A similar approach had led the ancient Greeks to associate certain polyhedra (the celebrated Platonic solids, each characterized by its own symmetry) with Aristotle's four elements or with planetary orbits.[41] Since then, mathematicians have perfected the notion of symmetry. Group theory deals with symmetries that are far richer and more complex than those encountered in ordinary geometry (as in the case of polyhedra). Modern particle physics uses these concepts in a very natural way, along the line followed by Kepler. The key to our modern successes is that the mathematical tools of our century, founded as they are on group theory, allow us to test the relevant ideas much more efficiently than was once the case. And yet, no one has ever been able to offer an a priori justification for such an approach, nor even to outline an intuitive rationalization. The advantage of mathematics in terms of efficiency unfortunately entails a loss in terms of immediate clarity. No physicist could possibly explain the gist of gauge theory, which is explicitly based on group theory, without resorting to complex mathematical derivations.

Harmony in Action

Although it may be somewhat oversimplified, one can draw a parallel between the search for harmony and the wave concept of matter. The latter runs diametrically counter to a corpuscular description of the world, which emphasizes the notion of particles or atoms. There is a clear-cut incompatibility between a corpuscule, a quasi-punctual object that can be assigned a precise location, and a wave, which by its very nature is "spread out," perhaps to the point of occupying all of space. The oscillatory nature of a wave conjures up the idea of harmony. It is something of a miracle that both views proved equally effective—the first in the fields of mechanics and thermodynamics, the second in the theory of light and electromagnetism.

In the seventeenth century, things were not nearly that simple. The synthesis achieved by Newton would not have been possible without Kepler's ideas about harmony, themselves reworked by Galileo. As it happened, this synthesis was founded on antithetical corpuscular and mechanistic worldviews. It is well known that Newton became quite taken with the idea of the harmony of nature, to such an extent that he devoted several years of his life to alchemy.[42] Newton's body of work is a story in paradoxes in that it encompassed in almost equal parts two decidedly antithetical views. Newton himself was acutely aware of it. In fact, it is precisely by not running away from these paradoxes that his true genius came through most vividly. Perhaps the best illustration is Newton's willingness to accept this mysterious action at a distance (which has come to be known as universal attraction), a concept "abhorrent to every mechanistic philosopher," for whom any interaction could only occur through direct contact.[43] Yet Newton embraced it, almost against his better judgment.

Newtonian mechanics was to subsequently enjoy spectacular successes. Even though it is basically a form of mechanistic atomism, from which the notion of harmony seems totally absent, its greatest triumphs would nevertheless result from its analytical content, specifically from what would come to be known—somewhat ironically—as "harmonic analysis." In the purest Newtonian tradition, analytical mechanics was to represent a partial return to a Cartesian approach. As it turns out, Descartes was one of the first individuals to introduce calculus into geometry. This union, which Newton rather frowned upon, was to subsequently prove remarkably fruitful.

Since that time, a harmonic view of things has resurfaced time and again even in theories closely allied with Newton's mechanistic atomism. "When I had the opportunity to ask professor Einstein how he had discovered relativity theory," recounts Hans Reichenbach, "he replied that it came to him because he was absolutely convinced of the harmony of the universe."[44] *In The World as I See it,* Einstein talks about a "cosmic religious feeling," which he sees as the "the strongest and noblest incitement to scientific research." Just what is this cosmic religion that seems such an integral part of the creative part of science? Frankly, Einstein himself is not quite sure. It does not involve any anthropomorphic god; it neither compels nor comforts. Its only indisputable characteristic is a desire "to experience the universe as a single significant whole."[45] One might also cite the case of the physicist Ernest Rutherford, who proposed in 1911 a planetary model that was essentially inspired by the astronomical structure and harmonic arrangement of the planets in the solar system.

One cannot escape the conclusion that a harmonic view of the world seems to consistently pervade physics. Even though it rarely competes openly with a purely mechanistic model, it quietly but steadfastly stimulates progress. Although a picture based on straight mechanism is adequate for any part of physics that does not require a quantum description, harmonic analysis remains at the ready in the wings even there, prepared to step in at a moment's notice. For instance, virtually no branch of physics nowadays can do without Fourier analysis, which is best described as a systematic search for mathematical harmony. The study of turbulence is a case in point. At first blush, it seems to belong in the field of hydrodynamics, a branch of physics that is fully consistent with a corpuscular view. But it soon becomes obvious that this approach is hopelessly inadequate to grapple with the complexity of the problem, and harmonic analysis must be called to the rescue in the form of classical Fourier analysis or any of its more modern equivalents, such as wavelet transform. These more sophisticated techniques are able to reveal discontinuities and singularities that correspond more or less to the intuitive notion of vortexes (which have nothing to do with Descartes's vortexes). It then becomes not only possible but indeed natural to first define and then exploit new concepts, such as scale invariance, catastrophes, and fractal structures, resulting from the synthesis of two opposing pictures—corpuscular and harmonic.

It is tempting to view quantum physics as the culmination of this type of dialectical process. The concept of wave function, seemingly halfway between a wave and a particle, would represent the best possible compromise integrating two competing models. Yet, quantum mechanics, which is essentially a mathematical description, has yet to find an interpretation everybody can agree on. Attempts keep seesawing, at least on a conceptual level, between a wave-like and a particle-like view. The true nature of reality seems to remain beyond our reach.[46] Perhaps the human mind is condemned to vacillate between two descriptions of the world fundamentally impossible to meld together. This balancing act may even be necessary for physics to move forward.

2

The Official Birth of Physics

The One, that is my true love. The One makes me free in servitude, fulfills me in adversity, enriches me in times of need, and gives me life in death.
—Giordano Bruno, The Infinite, the Universe, and the Worlds *(1584)*

The First Modern Thinkers

It is often said that physics came of age during the seventeenth century with such leading figures as Galileo, Descartes, and Newton. The previous chapter has already underscored this illustrious lineage. But it would be a gross oversight to ignore their many predecessors, whose own reflections exerted a strong influence from Greek antiquity to the Middle Ages. The transition from the ancient conceptions of nature, outlined at the beginning of this book, to a nascent science was a slow and laborious process, covering many pages in the history of ideas. Among the numerous shapers of the modern age, besides the Greek thinkers and mathematicians, are Thomas Aquinas (1228–1274), also known as St. Thomas, and Nicholas of Cusa (1401–1464).

We start with Thomas Aquinas because, until the Middle Ages, reason was entirely subordinated to faith, the nature of grace, philosophy, and theology. The world, tainted by original sin, was effectively despised and considered contingent. After rereading Aristotle, Thomas Aquinas contributed to the birth of a philosophy distinct from theology, indeed relatively autonomous from it. According to him, the purpose of philosophy was to bolster theology without becoming its servant. Because it aimed to understand the world and explain the essence of man, Thomist philosophy opened the door to unfettered speculation, hence to scientific knowledge. Thomas Aquinas's main thesis was that truth is revealed only through careful research, the validity of which can be gauged by the degree to which its conclusions conform to reality. Such an approach grants legitimacy to observation and experiment, in addition to syllogism, as valid paths to attain knowledge and objectivity: "When measured against the criteria of the human intellect, something will be deemed true if by its nature it is apt to produce a correct judgment about itself; on the contrary, those things whose nature is to appear different from what they are or are not will be held to be false."[1]

That is a primary criterion by which to judge any product of man's intellect. Although it may not be completely foolproof, it at least enabled theology to gradually find its own boundaries and fostered a separation of powers, as it were, enabling science to establish its own autonomy. Not until then could the concept of "laws of nature," and all their implications, begin to flourish. It is because there is this thing called *phusis*, with the inevitability of its laws, that science could be constructed as *logos*. In doing so, Thomas Aquinas minimized the temptation to glorify the forces of nature through beliefs steeped in the mysterious and to automatically invoke God's providence. As a result, an entire supranatural world, which had projected its shadow on all things, became blurred in man's psyche. From then on, nature was allowed to reveal itself in its raw reality. But its autonomy remained limited, for, as Thomas Aquinas reminded us, "to subtract something from the perfection of the creature is to restrict the very perfection of his creative power." In his view, the organization of the universe is the result of a transcendent divine will; the order of the world derives its existence and unity from divine simplicity since it emulates the plenitude God concentrates in himself by multiplying and duplicating it.

Nicholas of Cusa was another influential predecessor of Copernicus and Kepler. He displayed an uncannily "modern" mind in that he attributed a special role to the active workings of the intellect and favored a dialectical method that was open to the negative (almost an early prototype of Hegel's dialectics). He was able to articulate criteria defining the efficacy of knowledge building, and he did so in the face of a theology that had long obscured such criteria for its own benefit. In *Of Learned Ignorance*, published in 1439, Nicholas of Cusa explained that knowledge can be attained only through a process of separating and opposing phenomena, using words and concepts.[2] Yet a search for unity always underlies the object of knowledge. As a result, man finds himself trapped, so to speak, within the sphere of the intelligible, while at the same time he is consumed by a desire to inquire into what is unintelligible to him, such as the infinite and the divine, notions that are beyond the kind of oppositions normally apt to impart knowledge. That explains why there can be no awareness of God other than through a negative path, and why it makes sense to speak of negative knowledge (or "learned ignorance"), patterned after a negative theology, which holds that God makes himself known only as a reality that is beyond the human capacity to know.

Far from granting theology the high ground, this tack freed science from it. If God is incommensurate with our cognitive abilities, we cannot deduce from him all the things populating the universe and that we propose to make the object of science. As such, these things are strictly "casual," that is to say, contingent. Given our inability to derive them from a supreme unity, we are forced to try to understand them in and of themselves. All objects to be discovered must then be viewed as self-contained, and it is in this spirit that the human intellect embarks on its task to understand their intrinsic characteristics. In such a context, the absolute remains the ultimate goal of knowledge, but it resides within flawed creatures like ourselves and no longer remains locked beyond our grasp. As the sum of all finite objects accessible to man's intelligence, the infinite sketches out a distant outline of the length to which our knowledge can extend. Human beings are to be viewed in the image of the creator, and only to that extent do they have any hope of experiencing the divine.

Man's science can thus focus on the visible world without risking impiety or indignity, since in doing so it explores the invisible creator: "We can fathom the unity of unattainable truth through the vagaries of conjectures." Summing up, *Of Learned Ignorance* dispensed with theology while still preserving the uplifting character of the will to know. As far as Nicholas of Cusa was concerned, human knowledge is to the absolute what a polygon is to a circle. It is an approximation that mimics the real article only imperfectly.

In another of his books entitled *Dialogue on Genesis*, dated 1447, he took up the problem of the one and the multiple. Long before Kepler, Nicholas of Cusa advocated discarding the substantial or essential unity of the world and replacing it with the concept of harmony, according to which the world's unity would be based on proportions and even on proportions between proportions. In place of an abstract system of nature, he proposed a system made of concrete individual entities interrelated by laws. The purpose of knowledge, he maintained, is not to reveal nature or a being to us, but to lead us to laws and relations. He wrote: "Reason must study the physical world not by looking for types that do not exist, not by looking for them in a particular tangible phenomenon stripped of its individual accidents, but by establishing relations between phenomena." Such a conclusion is remarkably close to the premises of the modern scientific method. Even back then, Nicholas of Cusa insisted that science be formulated mathematically, particularly in the fields of cosmology, mechanics, physics, and chemistry (he even envisioned a mathematical medicine). We are witnessing nothing less than the initial stirrings of the scientific credo preached and practiced from the seventeenth century on.

Nicholas of Cusa's views were to echo in the writings of the physicist Joseph Fourier (1768–1830). In the early part of the nineteenth century, Fourier demonstrated that science does not have to look for the cause or the deeper nature of phenomena, but only for their laws. He was the first to work out a self-styled "analytical theory of heat," in which heat radiation, conduction, and convection were treated mathematically, without any hypothesis, mechanical or otherwise, about the nature of heat. Fourier's analytical theory does not involve any causal concept. Rather, it rests entirely on the assumption that phenomena are subject "to simple and constant laws, which may be discovered by observation, the study of them being the object of natural philosophy."[3] Fourier did not deny that causes exist, but merely that they are not essential for the purpose of developing scientific theories. His theory does not concern itself with the causes of heat, but only with its effects, which "constitute a special order of phenomena."[4] "To found the theory," added Fourier, "it was in the first place necessary to distinguish and define with precision the elementary properties which determine the action of heat."[5] Once catalogued, these phenomena can "resolve themselves into a very small number of general and simple facts; whereby every physical problem of this kind is brought back to an investigation of mathematical analysis."[6]

This development had a considerable intellectual impact for two major reasons. First, it introduced a new physical theory that was not wed to Newtonian mechanics. Second, it was the first instance of a rigorous science that did not resort to the notion of cause. Fourier himself had no philosophical agenda to promote. But his work impressed Auguste Comte and would later serve as a model to the positivists. From their perspective, Fourier's lesson was compelling: Mathematics proposes its own logic to nature; it introduces in a contingent way a degree of necessity in experimental facts. On the other hand, it tells us nothing about the real nature of things, being content to describe what can be observed.

Having said that, except for these few visionaries, physics did not truly exist before Galileo. The "orthodox" view of nature, inspired largely by Aristotle, was more akin to a natural philosophy based on sense perception. It was resolutely nonmathematical, as nature was primarily regarded as a process, or even as a giant organism. The world itself was thought of as a living biological entity (not unlike man) and, perhaps more importantly, teleologically driven. Indeed, the iconography of the Middle Ages often associated constellations or planets with specific parts of the human body. One of the most notable spokesmen for this movement was the Rosicrucian Robert Fludd (1574–1637), whom Kepler would later disparage for following the tradition of "alchemists, Hermeticists, and Paracelsists," in distinct contrast to a true mathematician like himself. Occultism, magic, superstition, astrology, and the like had not yet branched off from the mainstream of science and cannot be dissociated from the great humanist movement of the Renaissance.

A striking illustration of the then-prevailing attitude is given by Pierre Thuillier, who relates the moving words of an American Indian chief named Smohalla.[7] When Robert Boyle, an enthusiastic proponent of a mechanistic worldview, suggested to him that he should till his land and cultivate it rationally, Smohalla replied: "You urge me to till the land; am I to take a knife and tear at my mother's womb? If I do so, when I die, she will refuse to take me back that I may find my final rest. You ask me to dig into the ground and gather rocks; am I to forage under her skin and rip out her bones? If I do so, when I die, I will not be able to return to my body that I may be reborn. You encourage me to cut the grass and turn it to hay which I could sell to become rich like the white man. But how do I dare cut my mother's hair?"

Faced with the many inconsistencies of the world, the Aristotelians were well aware of the difficulty in discerning any real unity in it. Still, such a task was the duty of philosophers in Aristotle's opinion. In any event, Galileo overturned Aristotle by establishing the fundamental unification of physics. At about the same time, Descartes also claimed confidently to have unified the whole of science. Convinced that the principles he had isolated could explain everything by deduction, he purported to reveal "why there are mountains from which great flames sometimes erupt," and even why the blood of animals is red. By the time he was writing his *Principles of Philosophy*, he had become absolutely convinced he had found the key to unlock all the riddles of nature. In fairness, he did not expressly rule out the possibility that he may not have deciphered the true code of physics; but he had little doubt that he had done just that. The epitaph ordered by the French ambassador to Sweden for Descartes's temporary tomb attests to that confidence: "By linking the mysteries of nature with mathematical laws, he boldly hoped to unlock their respective enigmas with the same key." The use of a common key is all the more understandable since Descartes professed to have "reduced it [physics] to the laws of mathematics."[8]

Mathematics at the Service of Unification

Needless to say, mathematical views of the world—at least views inspired by certain aspects of mathematics—had been advanced before. Many centuries earlier, Euclid and Archimedes saw the world in terms of geometrical forms. The Pythagoreans and their followers were obsessed with relations between numbers. But such concepts had been dismissed by the largely dominant philosophy of the Aristotelians, who were unable to appreciate the full significance of mathematics. It would not be until the time of Galileo and Descartes that this discipline would attain preeminent status, preparing the way for Newton. Galileo decried the Aristotelians (among others) for their failure to take mathematics seriously. They may have understood the importance of experiments in constructing a worldview, but they were wrong in ignoring everything else. Galileo had the wisdom to envision a scientific method implying the predominance of reason over experience, the superiority of ideal models over empirical knowledge, the primacy of theory over mere facts. He was the first to proclaim that experiments and observation remain hollow and meaningless without the backing of a theory.[9] Science is not synonymous with experiment. Mathematics is required to formulate the questions one wishes to ask of nature, and mathematics—again and always—is the key to interpreting the answers nature is kind enough to offer.

The first unification achieved by Galileo was that of the method. By ascribing a pivotal role to mathematics, he provided physics with its underpinnings. Although this tradition has its roots with a number of thinkers in Greek antiquity, it failed to fully develop at the time. It finally blossomed in the seventeenth century, largely because physicists were by then primarily interested in mechanics, which is particularly well suited to mathematical treatments. One might argue, together with Antoine Cournot, that "Galileo created experimental physics and mathematical physics." He managed to "systematically capitalize on experimentation so as to force nature to lift the veil masking its secrets, to reveal the simple and fundamental mathematical laws lost to the weakness of our senses or obscured by the complexity of phenomena."[10] It is undeniable that physics is rooted in a mechanistic model. As soon as Newton added to it his own finishing touches, the model enjoyed such wide acceptance that, following Descartes's suggestions and in the wake of the enormous success of Newtonian physics, there was a rush to apply it to all other disciplines. It clearly acted as a guidepost for a broad range of scientific pursuits, to such an extent that some historians of knowledge do not hesitate to assert that Newton's natural philosophy inspired not only the mechanistic reduction of the living world to its physico-chemical substratum, but even Frederick Taylor's theory of work. One thing was beyond dispute, though, for both Galileo and Descartes: Unity is possible only through the bias of mathematics and, in Galileo's own words, "geometry [which] provides wings. Without its help, it is impossible to rise above the ground." According to Alexandre Koyré, "the essence of Galileo's scientific revolution boils down to the discovery of nature's language, to the realization that mathematics is the language of science."[11] With Galileo, a new conception of truth emerged, quite different from the old one in which it was subordinated to divine revelation. Truth was no longer bequeathed by God but became accessible to anyone willing to learn how to decipher its ways.

Descartes even explicitly articulated the "postulate of universal mathematism." The newfangled science aspired to be a mathematical translation of sensible phenomena. Mathematics was entrusted with the task of providing a unified vision; physics was reduced to geometry. What started as mere profession of faith with Plato (in *Timaeus*) became, from the seventeenth century on, the program of reason. That was the beginning of "the admirable science in which unite and culminate both the splendor of physico-mathematical cognition—for it is a universal triumph of mathematical clarity—and the splendor of spiritual interiority," as Jacques Maritain would later comment about Descartes.[12] And yet, Descartes was not quite the high point of this movement. That was not reached until Newton came along, even though he was opposed to numerous aspects of the Cartesian philosophy (atomism among them). Nevertheless, all these thinkers were in agreement about the vital role of mathematics.

Galileo, the Founder of Physics

The key to Galileo's success was in large part his ability to synthesize things, which he used to advance the cause of unification. His *Discourse on Bodies in Water* combined Archimedes' static approach with Aristotle's more dynamic treatment.[13] Starting from these two visions, each too restrictive on its own, Galileo forged a new theory that was far more unified. His efforts would in fact signal the birth of the physical sciences. History was to replicate this pattern time and again, weaving a web of multitiered unifications that were later confirmed.

The unity of physics is itself founded on mathematics. But an equally crucial ingredient is the notion of law. Indeed, what really propelled physics on its course and at the same time validated the concept of universe (the two go hand in hand, as we will see shortly) was the acceptance that some laws are "universal," meaning that they hold true everywhere, in all places and times, and for all objects. Thanks to these laws, we no longer have to think of the world as a morass of chaos and anarchy.[14] On the contrary, these laws unify the world and transform it into our universe by elevating it to the status of an object of science, the study of which becomes a legitimate endeavor. The invention of the universe came—Aristotle notwithstanding—with the unification of motion, of matter, and of space.

The Unification of Motion

Confronting questions of astronomy and cosmology, the founders of physics were primarily interested in mechanics and dynamics, two disciplines that were the first to benefit from the unification of their laws. "Astronomy, the science of the heavens, supplied the cosmic setting of the scientific revolution.... It was impossible for there to be a radical restructuring of natural philosophy in which mechanics, the science of motion, would remain unmoved, for motion plays a central role in any conception of nature."[15] For Galileo, "motion is the same as nothingness." What he meant was that motion (implicitly understood as uniform motion) and rest are to be considered equivalent; they are manifestations of one and the same concept, namely, inertia. This was a radical departure from the Aristotelian belief that natural motions are rectilinear (toward the bottom) here on earth and circular in the heavens.[16] After Galileo, physics was ready to state the principle of inertia, equally valid for circular and straight-line motions. Progress was further accelerated by Kepler's circuitous discovery of elliptical orbits. Even Copernicus had not dared challenge the preeminence of circles in the heavens. A few years later, Newton built on Galileo's work. He introduced the notion of force and finally brought to light the definitive formulation of dynamics. The process of unification associated with his name rests in large part on a series of mathematical calculations.

The Unification of Matter

For the Aristotelians, matter was divided in four elements subject to various combinations and associations. Any attempt to come up with a unified description of matter would have been doomed if for no other reason than that these elements had different destinies. Compounding the problem was the fact that such an element-based description was not meant to apply beyond the confines of our immediate surroundings. This restriction is one of the legacies of Aristotle. Indeed, he had made a fundamental distinction between our own world, which he perceived as imperfect and corruptible, and the distant world, which he believed to be composed of some unalterable "quintessence," making it perfect and incorruptible. By renouncing the Aristotelian vision and in effect introducing the concept of universe, Galileo's unorthodox posture opened to the door to, indeed demanded, unification.

In Galileo's view, our immediate surroundings could no longer be considered distinct from the outer world. When he trained his telescope toward the heavens, he discovered the ragged topography of the moon, "uneven, rough, and crowded with depressions and bulges. And it is like the face of the Earth itself, which is marked here and there with chains of mountains and depths of valleys."[17] Galileo concluded that matter must therefore be the same everywhere, just as "earth-like" on the moon or anywhere else as in our own backyard; it obeys the same rules, which means it is no less degradable in the heavens than on earth. Besides, Galileo objected to equating immutability with perfection because what is undegradable is not necessarily perfect: "Those who so greatly exalt incorruptibility, inalterability, etc., are reduced to talking this way, I believe, by their great desire to go on living, and by the terror they have of death.... There is no doubt whatever that the earth is more perfect the way it is, being alterable, changeable, etc., than it would be if it were a mass of stone or even a solid diamond, and extremely hard and invariant."[18]

There is nothing in Genesis to suggest that God used different materials to create earth and the heavens. Nonetheless, not until Galileo burst on the scene were these two worlds formally united in the minds of astronomers.

The Unification of Space

Admittedly, the distinction between sublunar and supralunar worlds had been called into question in the decades preceding Galileo's observations. Copernicus had already contested earth's privileged position, and Giordano Bruno had proclaimed that nature is everywhere of the same essence, made of the same matter. Tycho Brahe and Kepler went even further. They correctly surmised that the "new stars" they observed were located beyond the moon's orbit.[19] Even though they were obviously not eternal since they were "new," they belonged in the supralunar sphere, which was supposed to be immutable! To continue to claim that this world was unalterable had become untenable. Matter could no longer be considered in the Aristotelian image. What is more, both Tycho Brahe and Kepler saw comets whose orbits had to pierce through—nobody knew how—supposedly perfect crystalline spheres. These must therefore have been shattered to pieces, perhaps not literally since they did not truly exist, but figuratively speaking. All these observations awakened in both those modern thinkers deep misgivings about the Aristotelian worldview.

But it fell to Galileo to deliver the final blow to the traditional conceptions. Not only did he observe an earth-like moon with mountainous features, he also discovered that the sun, far from being perfect and immutable, is sometimes splattered with spots. No longer were the heavens the kingdom of order, perfection, and immutability. Galileo found that there is but one world with universal properties, and that was the universe itself! Confronted with this realization, he affirmed that the validity of physical laws is "universal." That conviction contained nothing less than the very foundation of all of physics, indeed of the whole of science.

This unification clashed with the prevailing consensus in two respects. If space is uniform, homogeneous, the same everywhere, then earth and man no longer occupy a privileged place in it. This spelled the end of anthropocentrism, which had already been battered by Copernicus, and even more so by Giordano Bruno, who would pay with his life for his revolutionary views. Likewise, by the wayside went Aristotle's theory of natural places. Aristotle had maintained that earth was the place of rest of heavy material bodies because they were spontaneously drawn to it by affinity; the heavens were the place of rest of fire, the element with an intrinsic character of lightness. Such distinctions became anachronistic. Celestial bodies were no longer different from their earthly counterparts, since the same laws applied everywhere. A short time later, Newton was to propose new laws that forever obliterated any such distinction.

The Birth of the Concept of Universe

There have always been men of the world. From now on, there will also be men of the universe.
—Paul Valéry

All these ideas were to be picked up again and elaborated upon in the twentieth century under the banner of the cosmological principle, which decrees as fundamental tenet that space is homogeneous, in other words, that all locations are identical.[20] The upshot was that the same laws apply everywhere. In short, physics is founded on the existence of a universe that is both unique by definition and unified by construction.

That it is unique is evidenced by the fact that all physicists devote their attention to the same one.[21] At least, that is what they all claim, and that claim becomes the springboard of their pursuits. Jacques Desmarets and Dominique Lambert have recently compiled a list of arguments showing the foolishness of the notion of multiple universes, on both physical and philosophical grounds.[22] The world is also unified, for all its parts—encircled as they are by space—are equivalent and obey the same laws. These laws qualify as universal precisely because they apply throughout the universe. It is impossible to conceive of the laws without the universe, or of the universe without the laws.

To be sure, we will see that physics will be able to make headway only by holding its methods in check, that is, by dissecting the universe in parts and compartmentalizing it in separate systems, in an approach we will describe as isolationist. But only to the extent that it can guarantee the universal applicability of its laws to any and all systems does physics claim its full legitimacy. In any event, the recognition of the universe marked the beginnings of physics. It is the common thread among all those who contributed to its birth, from Copernicus to Bruno and, later, Tycho Brahe and Kepler. Significantly, Kepler hailed Copernicus as the one who "destroyed solid orbits."[23] Not to forget Galileo, of course, nor Descartes who proclaimed that "matter in the heavens and on earth is one and the same"; and lastly, there was Newton, who elucidated all the properties of this new universe in terms of time and space.

This conception was hardly pleasing to the Church. In a universe where the laws of physics are everywhere the same as here, other worlds similar to ours (defined by earth and the solar system) are quite possible. Since God is not known for being lazy, he could not have been content creating a single one. For having expressed such views, Bruno was burned at the stake in the year 1600.

Until the early part of the twentieth century, science had no inkling that the universe could evolve at all. Even after discovering relativity, Einstein himself did not envision the possibility of a changing cosmos, so deeply rooted in everyone's mind was the belief in its immutability. The first models of the universe he proposed were entirely static. The Belgian physicist Georges Lemaître, a visionary in the purest sense of the word, was the first to theorize that things might be quite different. On the strength of observations and calculations based on relativity theory, he suggested in the 1920s that the universe might be expanding.[24] The speculation was soon borne out by measurements of the red shift of the emission spectrum of galaxies, which led to Hubble's law, established in 1929. Although it was originally conceived on a purely abstract and geometrical level, the expansion of the universe soon became the cornerstone of an entirely new brand of cosmology. Its physical implications, namely, that the content of the universe must be changing too, were far from immediately accepted. Once again, Lemaître was the first to come up with that notion when he proposed in the early 1930s his hypothesis of a "primordial atom," which at first left the scientific community rather skeptical. A few physicists came around in the ensuing years, but they had to contend with fierce and sometimes brutal resistance. Things settled down in 1965 when the big bang model became widely accepted after being spectacularly confirmed by the discovery of the diffuse cosmic background radiation.

The objections against the big bang model, which really amount to objections against the idea of an evolving universe, hark back to the opposition of the Catholic Church (at least of some of its members) to the new cosmology emerging in the sixteenth century. The big bang concept —indeed, the very idea of expansion—clashed with the centuries-old and deeply entrenched doctrine of a changeless universe. The reactionary trend against the notion of cosmic evolution was in a certain sense not unlike the position of the Aristotelians in Galileo's time. Founded on a secular myth, it deliberately ignored increasingly persuasive experimental observations. From a posture of reasonable prudence at first, it gradually degenerated into a stubbornly oppositional stance that grew more and more untenable in the face of mounting contrary evidence.

The most famous spokesman for that conservative school of thought was and remains without a doubt Fred Hoyle, to whom we owe the term "big bang"; he coined it as a scornful put-down targeted at Lemaître, whom he mocked as the "big bang man." In valiant attempts to preserve the myth of an eternal universe, he countered Lemaître's models with several of his own, some of which were actually quite ingenious. They were based on the unconditional constancy of the universe, accepted as an a priori premise in the form of a "perfect cosmological principle" that, at the cost of some modifications to the laws of physics, made it possible to construct so-called "stationary models of the universe."[25] All this seemed perfectly reasonable and scientifically sound, until things began to unravel as evidence that such models were inadequate to describe the cosmos accumulated.

The obsession to shore up models competing with the big bang theory, whatever the cost and despite their increasingly obvious shortcomings, caused the controversy to spread beyond the bounds of science. A few "die-hards of a stationary universe" continue to surface from time to time with bizarre pronouncements perpetuating a blind belief in a changeless universe.

The stance adopted by these anti-big bang crusaders is, in many respects, a throwback to the Inquisition three centuries earlier.

Fortunately, though, they do not enjoy the same clout and have to content themselves with a few publications largely relegated to the tabloid press. In Lemaître's time, they tried to exploit historical precedents by blaming his religious affiliation: Monsignor Lemaître happened to be a canon in the Catholic Church. The argument was specious at best. Unfortunately, it acquired an aura of respectability when Pope Pius XII gave Lemaître an awkward and unwelcome support, to the point that Lemaître felt compelled to ask the pontiff to refrain from intervening in this cosmological debate. Lemaître himself had always emphasized the physical, as opposed to metaphysical or religious, character of his cosmology.[26]

The need to respond to criticisms, as unjustified as they were, had at least one positive effect. It forced the experts to hone the scientific basis of their models, which in the meantime were found to be fully confirmed by experimental observations.[27] Yet, the fact remains that cosmology, like any other branch of physics, finds its inspiration and some of its justification in metaphysics. That is true of any theory or model, from Newton's gravitation to quantum physics, from thermodynamics to particle physics. Any scientific discipline carries deep within itself metaphysical roots.

Descartes's "Admirable Science"

The fields of dynamics and physics, invented by Galileo, spread and grew in stature, as the Aristotelian vision lost its hold. The year Galileo died, Newton was born in England. He pursued the work begun by his famed predecessor and developed a brand of physics that, for the most part, is still in wide use today. Newton's calculations promoted the advent of a unified physics. Yet it was Descartes (1596–1650) who articulated most clearly the need for such a unification.

Descartes dreamed of "blending the various mathematical sciences into a common science of proportions."[28] He believed harmony had to be expressed mathematically and science had to be unified. "All the sciences brought together are nothing but human wisdom, which is always one, always the same, no matter how varied the subjects to which it may be applied."[29] The diversity of objects that science deals with in no way dampens this aspiration toward unity, which is always fully justified. Science is one, because of the oneness of thought itself. This argues for the supremacy of mind over object, rather than the other way around.[30] Unity does not emanate from the diversity of impressions, but from the subject itself. To use Kant's phrase, it is embodied in the maxim "I think," which undergirds all our mental endeavors. For Descartes, the unity of science is thus a guarantee of quality. It is no accident that the original title of his *Discourse on the Method* was "Project of a Universal Science Destined to Raise Our Nature to Its Highest Degree of Perfection."

The first unity to guarantee is that of the method itself, as we have pointed out in our discussion of Galileo. Nothing could have been easier for Descartes, for whom science should be the work of a single individual. As he put it, "Often there is less perfection in works composed of several different pieces and made by different masters, than in those at which only one person has worked…just as it is indeed certain that the state of the true religion, the laws of which God alone has made, must be incomparably better ordered than all the others."[31] Descartes's ambition was a bit presumptuous and unrealistic: All of science worked out by a single man? True, there was no real scientific community to speak of yet. As it turned out, the one individual who came the closest to Descartes's ideal was Newton himself, as he did his work in virtually complete isolation. Be that as it may, the idea was now out in the open. Science can exist only if it is one, and it can be one only if it is built by a community with a common purpose. Descartes had correctly understood the source of science's strength: No matter how many different minds contribute to this great human enterprise, it manages to produce objective information recognized by all.

What kind of alchemy can create such a collective knowledge that—at least in its formulation—does not specifically acknowledge the individuals who contributed to it? This touches on the issue of the objectivity of scientific discourse. Descartes proposed a very simple explanation: "Since reason taken in the state of nature is the sufficient instrument of all knowledge, reason on the other hand being perfectly one in each man, why should science not be one, why should it not be of the very unity of the human mind?…It is in the unity of the admirable science, which will imply one single light and single mode of certitude, that 'the Spirit of Truth' opens to the philosopher the treasure of all the sciences."[32] The path Descartes saw for himself was clearly marked: It is every philosopher's duty to unmask and reveal science in all its beauty, continuity, and unity. Much later, in 1844, Auguste Comte would reiterate the need for this type of intellectual communion by stating that "science fulfills the desire to unify the knowledge of all mankind."[33]

Isaac Newton, a Man of Profound Syntheses

The path opened by Kepler and furthered by Galileo (not to overlook the contributions of Boyle, Hooke, and many others) culminated in the Newtonian revolution, which many consider the real beginning of physics characterized by a systematic program of unification. The historian Richard Westfall identified four ingredients in this Newtonian unification: Mathematics, optics, and mechanics, and gravitation, although the latter properly belongs in the province of mechanics. In a span of fifteen years—between 1672 and the publication of his *Principia*—Newton caused a radical change in our conception of the world. He did so by amalgamating the many rival systems competing up to that point into a single one.

The starting point was a synthesis of Descartes's mechanistic physics with Gassendi's atomistic philosophy. The synthesis was accomplished first for matter, and subsequently for light.

Next came the synthesis of different types of motion. Already begun by Kepler and Galileo, it was formally incorporated by Newton in his own theory. We have already mentioned the merging of rectilinear and circular motions. But it was the whole Aristotelian concept of movement that was being shattered. Westfall put it this way: "To Aristotle, to move was to be moved. The motion of any body required a moving agent."[34] Newton introduced the notion of inertia, which allowed motion without cause. Aristotle had made a distinction between "natural" motion and "violent" motion. This dichotomy had come under assault with Galileo and was forever destroyed by Newton. Motion became synonymous with ordinary displacement: "Motion is simply a state in which a body finds itself, a state to which it is indifferent."[35] With Aristotle, any motion implied an ontological change, rather like the growth of a tree. In the end, what Newton built was a genuine science of motion, to which he gave a mathematical foundation.

Finally, Newton unified the concept of matter. From that point on, it became almost universally accepted "that physical nature is composed of one common matter, qualitatively neutral and differentiated solely by the size, shape, and motion of the particles into which it is divided. All agreed that the program of natural philosophy lay in demonstrating that the phenomena of nature are produced by the mutual interplay of material particles which act on each another by direct contact alone."[36]

These are the essential components of a mechanistic philosophy that all worldviews were built on in the seventeenth century. Until Newton came along, Descartes was one of its most effective proponents. His theory of vortexes fit right in, although it later proved incorrect. Newton was staunchly opposed to it, but overall he followed very much the same line of thinking. The one fundamental exception was his notion of action at a distance. The seventeenth century did away with Aristotle's philosophy of nature, replacing it "with a new philosophy for which the machine, not the organism, was the dominant analogy."[37]

The unification of matter also signaled the birth of a new chemistry.

Strictly speaking, there never was such a thing as an Aristotelian chemistry, even though Paracelsus had proposed in the sixteenth century to add to Aristotle's four elements three so-called "principles"—"salt, sulfur, and mercury, which functioned as the primary agents in chemical explanations."[38] But they amounted more to "a doctrine of substance than a chemical theory." Indeed, in keeping with an animistic view, Paracelsus's three principles were supposed to represent body, soul, and spirit. There came a time when Newton himself—that is another huge paradox—was literally enthralled by alchemy, which was based on just such an animistic conception of nature.

Beginning in 1661, Robert Boyle, the generally acknowledged "father of chemistry," took Aristotle and Paracelsus to task and introduced the idea of a mechanistic chemistry. He insisted that "elements and principles do not exist. What does exist is the qualitatively neutral matter of the mechanical philosophy, divided into particles differentiated only by size, shape, and motion. From their various combinations arise all the appearances of the substances with which chemists deal."[39] In Westfall's opinion, that was a turning point: "To chemists Boyle offered full participation in the fraternity of natural philosophers. By mechanizing chemistry, he effectively obliterated the barriers that had separated their enterprise from the rest of natural philosophy."[40] It marked the beginning of chemistry as we know it, made possible by the notion that matter is one.

This series of syntheses, consolidations, and unifications marked the beginnings of physics. Exit the monistic attempts of the Greeks; exits the picture of a living organism. A new model was adopted, based on the concept of a universe patterned after a machine. It was particularly well suited to a mathematical description, which would make it at least more operational, if not more comprehensible.

As a result of this unification, the universe acquired a brand-new look, one that was hardly recognizable. As Alexandre Koyré observed, the process "occurred by substituting for our world of quality and sense perception, the world in which we live, love, and die, another world—the world of quantity, of reified geometry, a world in which, though there is a place for everything, there is no place for man."[41]

3

The History of Physics
A Series of Unifications

The world is a complicated place.
—Steven Weinberg, Dreams of a Final Theory

The Quest for Unity

The search for unity did more than just bring about the birth of physics. It has remained at the forefront of physicists' preoccupations ever since. Perhaps it is only a dream, but it has nevertheless captured an enormous amount of attention during the course of the centuries, and often with dramatic results. The few examples we are about to review in this chapter should inspire us to be more circumspect about the current state of physics and its prospects for a more profound unification.

The rise of modern physics is best summed up as a series of unifications. Some key events along the way include the Newtonian synthesis of atomism and mathematism; the unification of space, which led to the concept of universe; the consolidation of astronomy and physics; the unified conceptions of matter and of different types of motion. Many subsequent advances also qualify as successful unifications.

During the second half of the nineteenth century, Maxwell successfully lumped electricity and magnetism together into a new theory called electromagnetism. He went on to further unify electromagnetism and optics by demonstrating that light is a just particular kind of electromagnetic radiation. At the beginning of the twentieth century, the theories of special and general relativity prompted a whole array of new unifications: electromagnetism with kinematics and, ultimately, dynamics; space with time; and matter with radiation (themselves related to space-time through the bias of gravitation).

Also in the twentieth century, quantum physics provided yet another synthesized vision of matter and radiation in the context of particles and waves. Later, quantum field theory integrated quantum mechanics with special relativity; the key to this synthesis was a unified treatment of matter and its interactions. In turn, this way of looking at things opened the door in the second half of the century to further unifications, of which the best-known prototype is the electroweak theory.

Finally, the science of cosmology, which is in the midst of rapid developments, is unifying by its very essence, since it considers the universe in its totality, encompassing—by definition—all branches of physics.

Light and Electromagnetism

Arguments about whether light and matter are best described as waves or particles date back to antiquity. Descartes, who was the first to correctly state the laws of reflection and refraction, thought of light as tiny propagating particles. Despite his own discovery of the spectral decomposition—hence harmonic nature—of light, Newton concurred with Descartes's view.

In the opposite camp, Christian Huygens (1629–1695) was already comparing the propagation of light to that of waves on the surface of water. What medium—tentatively called "ether"—might be playing the role of water in the world of light? That remained a profound mystery. But the wave picture enabled Huygens to explain not only reflection and refraction, but also the phenomenon of birefringence discovered in 1669 by Erasmus Bartholin (1625–1698). The nineteenth century brought spectacular confirmations that light behaves as a wave. In 1801, the English physicist Thomas Young (1773–1829) observed interference phenomena; in 1810, Etienne Louis Malus (1775–1812) discovered the polarization of light. A few years later, the French physicist Augustin Fresnel (1788–1827) completed his own experiments and interpreted them correctly in terms of waves. It looked increasingly as though that was the only way to account for all these phenomena. After Fresnel's death, numerous other experiments further confirmed the wave nature of light, notably those of Armand Hyppolite Fizeau (1819–1896) and

Léon Foucault (1819–1868), who demonstrated in 1850 that the speed of light was lower in water than in vacuum.[1] The concept of light as a particle was essentially abandoned from that point on.

In the early part of the twentieth century, the wave-like nature of light was called in question all over again in the wake of the 1887 discovery of the photoelectric effect by Heinrich Hertz (1857–1894). As Albert Einstein showed, the effect could not be explained on the basis of waves. Shortly thereafter, an understanding of the Compton effect also required a particle-like description. The phenomenon involved collisions between light particles, or photons, and electrons. The sticky point was that light obviously had to be the same regardless of the type of experiment. Neither a wave model nor a particle model could explain, by itself, all known experimental results. A new way of looking at things was called for. Was it necessary to accept, as Louis de Broglie (1892–1987) had first suggested, that "the wave and particle aspects of light are two complementary sides of the same reality"?[2] Quantum mechanics, which we will discuss shortly, would finally succeed in providing a unified treatment capable of accounting for this duality.

When Hans Christian Orstedt (1777–1851), then professor at the University of Copenhagen, announced in 1820 that he had just observed that a wire carrying an electrical current causes a magnetized needle placed close by to deviate, he was probably completely unaware that he had just inaugurated electromagnetism, which was to become one of the foundations of nineteenth-century physics. Granted, a probable connection between electrical and magnetic phenomena had been suspected for some time. It was well known, for instance, that storms, which Benjamin Franklin had previously demonstrated to be electrical in nature, could cause a compass to go haywire. But no one had ever been able to measure such phenomena with any precision or reproducibility.

In the early part of the nineteenth century, a crude theoretical understanding of such phenomena rested on two then quite distinct pillars—electrostatics, which had been developed by Charles Coulomb (1736–1806) and Denis Poisson (1781–1840), and magnetostatics. Electrostatics described interactions between electrically charged objects, while magnetostatics dealt with those between magnetized bodies. There were certain similarities between the two, for example, the fact that objects can attract or repel one another. But, for the most part, the two classes of phenomena appeared to be of a completely different nature. A magnet and an electrically charged object have no influence on each other when both are at rest relative to each other; an electrical charge is either positive or negative, while a magnet always has two opposite poles, even after being broken in two pieces.

One week after Orstedt's announcement, André-Marie Ampère (1775–1836) proposed an explanation. To analyze the problem, he considered infinitesimally small filaments of a conducting wire carrying an electrical current and asked himself how such filaments acted on one another. He realized that the magnetic problem could be reduced to electrical interactions between two separate wires. The reason a wire can act on a magnet is that a magnet is itself equivalent to a multitude of elementary current loops. Ampère had just found the key to understanding all magnetic phenomena. He had put his finger on the link between electricity and magnetism. Magnetism simply reflects the existence of electrical currents or, more generally, of charges in motion.

This explanation was corroborated by the work of Michael Faraday (1791–1867). Fascinated by Orstedt's experiments, Faraday set out to demonstrate the reverse effect, which is the creation of an electrical current in a conducting wire under the influence of a magnet. He finally succeeded in 1831, after observing that the desired effect was produced only if the magnet was moving in relation to the wire (an effect now known as electromagnetic induction).

In the process, Faraday reawakened an age-old controversy in physics, centering on the issue of instantaneous action at a distance. The notion that any interaction between two objects depends only on the nature of these objects and their distance, and not at all on the medium between them, goes against the grain since it says nothing about how the interaction propagates from one object to the other. Newton's gravitation relied on precisely such a notion; so did electrostatics and magnetostatics, even after the latter had been revamped by Ampère. Faraday favored propagation through a medium. After all, such a picture was supported by an experimental result Faraday himself had established. He had proven that the amount of charges accumulating on two conducting plates separated by an insulator depends not only on the distance between the two plates, but also on the nature of the material between them.

Unfortunately, Faraday's proficiency in mathematics was a bit limited, and he was unable to prove his intuition rigorously. That task was begun by William Thomson (1824–1907) and completed by James Clerk Maxwell (1831–1879). In order to describe how electrical charges, whether fixed or in motion, affect the surrounding space, Maxwell developed the concepts of electrical and magnetic fields, which characterize what one might call the "electromagnetic state" of any point in space. Drawing on the mathematical formalism traditionally used to describe the propagation of perturbations, namely, partial differential equations (the very same ones Fourier had used earlier in the century to model the propagation of heat), Faraday reworked and refined the laws of electromagnetism as they were known at the time. In 1864, he arrived at nine fundamental equations (which have since been reduced to four). His exceedingly elegant theory made it possible—and that was a crucial result—to calculate the propagation velocity of electrical and magnetic perturbations. This velocity turned out to be identical with the speed of light, quite an unexpected result considering the special status light was accorded at the time.[3] Maxwell never went the extra mile by recognizing that light is in fact an electromagnetic wave in its own right. Instead, he simply concluded that the propagation of both electromagnetic waves and light results from the vibration of the same hypothetical medium that had served him so well in deriving his famous equations. Maxwell was, of course, referring to ether, a medium so subtle that no one had ever detected it.

It fell to Heinrich Hertz (1857–1894), a student of Hermann von Helmholtz (1821–1894), one of the great physicists of the day, to consolidate the triumph of Maxwell's theory. He began by ridding the theory of its most objectionable mechanical ingredients. He discarded the ether altogether, retaining only electric and magnetic fields, which saw themselves promoted from convenient computational aids to legitimate physical entities. He proceeded next to verify the fundamental prediction of Maxwell's theory, which is that electromagnetic waves propagate at the speed of light. With an electrical device of his own design, in 1887 Hertz produced electromagnetic waves with a long wavelength—what we would now call Hertzian waves. He managed to measure their propagation velocity and confirmed that it was indeed the same as the speed of light. He carried out further experiments demonstrating that these waves could be reflected and refracted in exactly the same way as light.

By then, there could no longer be any doubt that there was in fact no difference between the two. The electromagnetic nature of light was firmly established, giving Maxwell's equations far more generality than initially recognized.[4] A set of just four relatively simple equations had the power to unify not only electricity and magnetism, but the entire field of optics as well. Never before in the history of physics had such a restricted number of laws been able to account for such a wide variety of phenomena.

Relativistic Unifications

Special Relativity

Nineteenth-century physics was rooted in Newtonian mechanics, which deals with the motion of material objects, and electromagnetism, which is the science of light and all other electromagnetic phenomena. For a long time, each of these theories seemed to be perfectly correct in its own domain. Gradually, though, evidence began to emerge that they actually contradicted each other.

Mechanics is entirely founded on the principle of relativity, which was originally stated not by Albert Einstein, contrary to what many people believe, but by Galileo around the year 1600. By virtue of this principle, events unfold in exactly the same way in an airplane—Galileo used the example of a ship—whether it is flying at cruising speed or sitting idle on the tarmac. For instance, if a flight attendant drops a glass of water, it falls along a precisely similar path in the "Galilean frame of reference" attached to the plane as it would in a restaurant on the ground. No physics experiment of any kind can be devised to determine whether the plane is in the air or waiting at the gate, at least as long as its motion takes place at constant speed and in a straight line. As Galileo himself observed, the motion of the plane—or of the ship—is "as though it were null"; it doesn't count since it cannot be felt.

What are the consequences of this principle? While in the air, the seated passengers are (hopefully) not moving with respect to the walls of the aircraft; but they are (no less hopefully) moving relative to the earth, which is itself moving relative to the sun, which in turn is moving relative to the center of our galaxy, which itself is not standing still either. In short, the principle of relativity asserts that nothing is at absolute rest.

Now, the theory of electromagnetism, as worked out by Maxwell, tells us that light is a wave. For a nineteenth-century physicist, a wave was a disturbance that propagates by causing something in the medium in which it travels to vibrate. In the case of an ocean wave, the archetype of an oscillatory phenomenon, what vibrates is the water or, more precisely, its surface. For sound waves, it is the air (which is why sound cannot propagate in vacuum). In the case of light, the seat of vibrations is the "ether," or so everyone believed in the nineteenth century.

We have to imagine, along with Maxwell, that the world is filled, down to its tiniest nooks and crannies, with this medium called ether, the existence of which is necessary for light to propagate. What is this ether made of? What does it look like? Is it like water, air, glass? Does it have weight? Is it a solid, a liquid? Is it elastic? Maxwell's theory provided a few tentative answers, but they were invariably murky, claiming that ether is "almost certainly" colorless, "probably" weightless, and so forth. In point of fact, as time went on, the ether progressively lost all the physical properties that had traditionally been attributed to it, until a single one was left: It had to be in a state of absolute rest. The trouble was that this was in complete contradiction with the principle of relativity.

Hence a very thorny dilemma. If one accepts Maxwell's theory, which is perfectly consistent with experimental evidence, one is forced to accept that light propagates in an absolutely immobile ether; this is tantamount to abandoning the principle of relativity, and therefore to shooting down classical mechanics. If, on the other hand, one takes the principle of relativity seriously, one is forced to discard the ether. But in that case, how can light propagate?

Einstein eventually resolved this contradiction in his famous 1905 paper that unveiled relativity theory in the modern sense of the term. He began by declaring the ether dead. Light does not require the vibration of any physical medium at all. It propagates even in vacuum. Einstein next took one of the implicit consequences of Maxwell's theory and elevated it to the status of principle. It states that light must *always* travel at the same speed *c*, equal to 300,000 km/s, regardless of any possible motion of an observer. When a car approaches me—the observer—with its lights turned on, the light it emits propagates with the same speed relative to me as if the car were standing still. The classical law of velocity addition, as it was originally stated by Galileo, does not apply to light. The speed of light relative to me is not equal to the sum of its own propagation velocity and the speed of the car, contrary to intuition. Einstein's relativity theory introduced some modifications to the laws of addition of velocities so as to make them consistent with the postulate of invariance of the speed of light in all Galilean reference frames (assumed to be in uniform and rectilinear motion with respect to one another).

The ether violated the principle of relativity, which precluded anything from being in a state of absolute rest. Getting rid of the ether was a welcome relief in that it reinstated relativity. Further adding the postulate that light propagates at the same speed *c* relative to any observer, whether in uniform motion or at rest, required rebuilding physics from the ground up. That is precisely what Einstein did.

In order to provide a footing to this new physics, it was necessary to introduce what was then a revolutionary concept of space-time, replacing the traditional notions of space and time that had been, up to that point, completely separate. The hyphen in the word space-time is a constant reminder that the two concepts are to be thereafter inextricably linked. One remarkable consequence of this new picture is that neither dimensions nor durations are absolute quantities any longer. Instead, they depend on the frame of reference in which they are being measured.

To gain a better appreciation for why that is, consider two frames of reference R and R', both of which we assume to be Galilean. The coordinates defining an event in the frame R are denoted (x,y,z,t). The first three (x,y,z) represent the location where the event occurs, while the fourth (t) specifies the time of occurrence. The same event in frame R' is recorded as (x',y',z',t').

Assume next that R' is in uniform motion with respect to R along the x-axis, with a velocity *v*. The coordinates of an event measured in frame *R*' can be calculated from the coordinates of the same event measured in *R* through what is called a Lorentz transformation, the expression for which is:

$$x' = \gamma(x - \beta ct)$$
$$y' = y$$
$$z' = z$$
$$t' = \gamma(t - \beta x/c)$$

In these equations, the quantity β is defined as the ratio v/c, and the coefficient γ is equal to $1/(1 - \beta^2)^{1/2}$. Since the velocity v can never exceed the speed of light *c*, the quantity β is restricted to values less than 1, while γ is always greater than 1.

Consider first two events that take place in two different locations in the frame of reference *R*, but that are simultaneous in the sense that they occur at the same instant t. Their coordinates in frame *R* are given by (x_1,y_1,z_1,t) and (x_2,y_2,z_2,t). The question is: Are they also simultaneous in frame *R*'? Using the transformation equations written above, the two events occur in *R*' at instants t'_1 and t'_2, given by:

$$t'_1 = \gamma(t - \beta x_1/c)$$
$$t'_2 = \gamma(t - \beta x_2/c)$$

It is immediately obvious that if x_1 and x_2 are different, t'_1 and t'_2 cannot be equal (as long as β is not equal to zero, which would mean that the two frames of reference are standing still relative to each other). The inescapable conclusion is that two events simultaneous in *R* are not simultaneous in *R*'. It follows that the very concept of simultaneity is no longer absolute, in total contradiction with Newtonian physics. That was a stunning outcome of relativity theory.

Furthermore, with the expressions given above, it is easy to see that the length of a ruler is not the same when measured in *R* and *R*'. Likewise, it is straightforward to show that the time interval between two events depends on which frame of reference it is measured in.

However, let us turn our attention to what is called the space-time interval Δs separating two events indicated by the subscripts 1 and 2. This quantity is defined in frame *R* by:

$$(\Delta s)^2 = (x_2 - x_1)^2 + (y_2 - y_1)^2 + (z_2 - z_1)^2 - c(t_2 - t_1)^2$$

The same interval measured in frame *R*' is given by:

$$(\Delta s')^2 = (x'_2 - x'_1)^2 + (y'_2 - y'_1)^2 + (z'_2 - z'_1)^2 - c(t'_2 - t'_1)^2$$

Utilizing the expressions for the Lorentz transformations, it is elementary to show that $(\Delta s')^2$ and $(\Delta s)^2$ are precisely equal. That result is crucial. It means that the space-time interval between two events is independent of the Galilean frame of reference in which it is calculated. The quantity Δs is said to be a "Lorentz invariant."

It has often been pointed out that relativity theory is really a misnomer. Indeed, it might have been more appropriate to call it "theory of invariants," since it focuses on what does not change regardless of the observer's point of view. In that sense, it is structurally unifying.

The invariance of space-time intervals explains why a clock in motion relative to an observer always runs more slowly than a clock at rest. The closer to the speed of light it moves, the slower it runs. This perspective effect in space-time is called "time dilation."

On September 27, 1905, Einstein sent out a sequel to his June paper. It included the formula $E = mc^2$ with which his genius is to be forever associated, which came to symbolize all of relativity theory. The equation expresses the fact that the energy content of a body is related to its mass. Even when at rest, a body has an internal energy proportional to the quantity of matter it contains. Rest mass and internal energy become entirely equivalent. As it happens, the orders of magnitude involved are simply mind-boggling. For instance, a half-pound bar of butter (or of anything else with the same mass) corresponds to an energy of about 2×10^{16} joules! This is a huge number compared, for instance, with a kinetic energy of only about 0.11 joule if that same bar were to fall at a speed of 1 meter per second. It explains why this all-important equivalence between mass and energy was not discovered until 1905. The energy that can be imparted to an ordinary piece of matter by accelerating it through ordinary means is completely negligible in comparison to its enormous internal energy content.

One of the consequences of Einstein's formula is that rest mass can be converted to other forms of energy. That is precisely what goes on in nuclear reactions. Conversely, kinetic energy (related to the speed of an object in motion) can be transformed into rest mass; in other words, it can be materialized. That is the principle behind the creation of elementary particles in accelerators.

To summarize, special relativity was introduced with a specific unifying purpose in mind. Electromagnetism was characterized by a finite propagation velocity c of light that, strangely, seemed to be invariant regardless of the circumstances. By contrast, the dynamics of phenomena involving matter, such as Newtonian physics described it at the turn of the previous century, obeyed a set of laws prescribing how velocities combined. These laws could in principle lead to velocities exceeding that of light and certainly did not require the latter to be invariant.

Interestingly, this description was perfectly satisfactory within its own field, namely, the dynamics of material bodies. There was not a single shred of experimental evidence suggesting any need to revise the conventional view. The problem came about purely as a result of theoretical efforts aimed at devising a common and unified understanding of electromagnetism and kinematics. Only then did the inherent contradiction become apparent. From this perspective, relativity is truly the offspring of theory.

The solution proposed by Einstein actually built on the earlier work of Hendrick Lorentz (1853–1928) and Henri Poincaré (1854–1912). In the end, space, time, and motion turned out to be the same for material bodies as for optics and electromagnetism—the same spatial and temporal symmetries applied throughout.

The unifying character of relativity theory went even further than originally intended. Through the work of Hermann Minkowski (1864–1909), it would lead to a synthesis of the notions of space and time, which came to be integrated into this then-strange concept of space-time. Special relativity allows many aspects of kinematic theory to be viewed as resulting from pure geometrical invariances, which generalize in a space with four dimensions (space + time) the familiar invariances with respect to rotations and translations in ordinary three-dimensional space. In a four-dimensional space-time frame of reference, the Lorentz transformations play the role of ordinary rotations and translations.

General Relativity

Special relativity had dethroned space and time viewed as absolute concepts. It had rebuilt in Minkowski's space a completely new mechanics based on the premise that the speed of light is the same in the frame of reference of any observer, making this quantity a fundamental constant of physics. Newtonian mechanics, on the other hand, was based squarely on a postulate that no one had ever been able to verify directly. The postulate stated that the effect of gravitation between two bodies A and B, which manifests itself as a mutual attraction expressed by Newton's law, propagates instantaneously through space. If A were to suddenly change shape (for instance, from a sphere to a disk), B would immediately be "aware" of it, regardless of the distance between them.

That was a troublesome contradiction with the new relativity theory.

After searching in vain for a solution over a period of ten years, Einstein eventually found himself forced to radically alter the very concept of gravitation such as Newton and his followers had understood it. In 1915, Einstein proposed what came to be known as the *general theory of relativity*. The name itself reflected the fact that the theory resulted from a generalization of notions previously developed in the context of special relativity and extended to gravitation. General relativity really amounts to a new theory of gravitation. A number of alternative theories have been proposed since then, none of which have so far been confirmed by tests, no matter how precise. But they all follow closely Einstein's line of thinking, and measurements suggest that differences with Einstein's own theory, if any, must be very small.

In the mid-1850s, the German mathematician Bernhard Riemann (1826–1866) was busy extending the work of the Russian Nicolay Lobachevsky (1792–1856), who had been interested in non-Euclidian spaces characterized by a geometry that is quite different from that of our ordinary space. In 1854, Riemann, then twenty-eight years old, suggested that gravitation may not be a "true" force at all, but simply a manifestation of the curvature of space. Einstein picked up on this idea and developed it to its full potential. The geometry of space-time, which had remained familiarly flat in special relativity, was to be viewed from then on as curved, owing to the masses in its midst. The motions of earth or any other planet no longer resulted from some instantaneous action at a distance of Newton's gravitational force. Rather, the trajectory of such objects was simply determined by "distortions" in the fabric of space-time, imposed by the presence of the massive sun. As the American physicist John Wheeler put it, "curvature tells matter how to move, and matter tells space-time how to curve."

This totally new perspective would have remained an interesting curiosity, without any practical consequence, were it not for two crucial experimental tests. One would bring Einstein instant fame, and the other would resolve once and for all a puzzle that had long confounded astronomers.

The first test had to do with the way light behaves when it passes close to a massive object. Newton, who believed that light was made of particles, had suggested that it would follow a path bent by the gravitation due to the large body, much like the deflection experienced by any other projectile, regardless of its mass. But verifying such a hypothesis was fraught with enormous difficulties. No one had even tried it. As it happens, Einstein's new relativity theory predicted a deflection exactly twice as large as that predicted classically. While observing a total solar eclipse in 1919, the astronomer Arthur Eddington (1882–1944) carried out measurements that turned out to agree precisely with Einstein's predictions. Since then, many additional confirmations of this effect have been brought about by the spectacular development of the field of gravitational optics. We now have a fairly detailed understanding of the circuitous path of light as it wanders across the universe, skimming over various masses in its long journey. The resulting mirages and gravitational lens effects have become the delight of astrophysicists.

The second test concerned itself with the orientation of the plane of the ellipse described by a planet as it moves around the sun. According to Newtonian mechanics, the orbital plane must remain fixed in relation to distant stars (since angular momentum of the sun-planet system is to be conserved). As the precision of measurements improved over time, nineteenth-century astronomers began to notice that the plane of Mercury's orbit around the sun precesses at a rate of 43 angular seconds per century. Such a result was totally puzzling in the context of classical mechanics. But general relativity was able to account for it down to its minutest details, as it did for similar precessions observed in double-star systems.

Since Newtonian mechanics remains entirely adequate for calculating the trajectory of a rocket or a planet, there must exist some criterion defining its range of validity. One such criterion has indeed been worked out. It is analogous to the coefficient β introduced in special relativity, which compares the velocity of the object of interest to the speed of light. Similarly, in the present case, there is a coefficient α that quantifies the relative importance of the presence of a mass in space-time. Consider a mass M contained within a sphere of radius R, and a small mass m on the surface of that sphere. The coefficient α is defined as the ratio of the potential energy of m under the influence of the gravitation of M (the energy that causes m to fall toward M) and the rest mass mc^2 of m. In mathematical terms, this is written as:

$$\alpha = (GMm/R)/(mc^2) = GM/Rc^2$$

where the quantity G is the gravitational constant. The reader will have noticed in the above equation that the small mass m drops out. In other words, the coefficient α is independent of m; it depends only on the structure of the space-time associated with M, as it makes its presence known to m. Moreover, the quantity $v_L^2 = (GMm/R)$ turns out to be the square of the escape velocity required for m to break free of the gravitational pull of M. The closer this velocity is to the speed of light (either because M is large or R is small, or both), the closer to 1 the value of α becomes. The criterion we are looking for becomes a matter of comparing the value of α to the number 1.

At the surface of earth, for example, α is quite small, being of the order of only 10^{-9}, which explains why Newtonian mechanics is perfectly accurate in our ordinary world. At the surface of the sun, α is already about 10^{-6}, in which case relativistic effects begin to be detectable, as Eddington found out in 1919. But at the surface of massive objects such as neutron stars, in which a mass comparable to that of the sun (about 10^{30} kg) has collapsed into a sphere whose radius is no more than a few kilometers, α may be as large as 0.1, and deviations from Newtonian mechanics become quite substantial. Things get even wilder in the vicinity of a black hole, in which case α can approach unity.

General relativity is expressed in a mathematical language that is not nearly as easy to handle as in Newton's theory. Where ordinary vectors once sufficed to describe forces and accelerations, one is now forced to resort to mathematical objects called tensors. With ten components each, these entities are considerably less friendly to manipulate. One such tensor describes space-time curved by the influence of all the masses involved. Another describes the energy and momentum of all the matter present in that space-time. Einstein's equations relate these two tensors and make it possible, at least in principle, to calculate simultaneously the curvature of space and all the possible motions it can support. Unfortunately, the calculations, which involve a total of ten equations, can be exceedingly complex.

General relativity is based on a mathematical expression of the principle of equivalence. Translated loosely in words, this principle states that the motion of a particle in a gravitational field is independent of the internal structure or composition of that particle. Legend has it that Galileo tried to verify this principle by dropping various objects from the top of the leaning tower of Pisa, although the story is probably just that—a legend. For Einstein, who viewed gravitation as a purely geometrical effect, a rigorous formulation of the principle went something like this: All laws of physics, as well as its fundamental constants, are locally identical, whether in the presence or absence of gravitation. No observer confined to a windowless elevator in free fall or in a spacecraft with its engines shut down can conduct any experiment whatsoever that would reveal to him the masses toward which he is falling.

By encompassing and extending the range of applicability of special relativity, general relativity constitutes a bold set of successful unifications. In Niels Bohr's words, "Einstein succeeded in remolding and generalizing the whole edifice of classical physics and in lending to our world picture a unity surpassing all previous expectations."[5]

The concept of space-time, already introduced in special relativity, lumped space and time together. From that point on, this geometrical entity (more precisely, its metric) acquired an almost concrete explicative power, to such an extent that the physicist Jean-Marie Souriau does not hesitate to accord it a status comparable to that of the elusive ether, which physicists had pursued literally for centuries. With space and time henceforth united, the new theory provided a common descriptive framework for kinematics, dynamics, and optics. Matter and radiation propagate according to precisely the same laws and are furthermore considered two equivalent forms of energy.

The equivalence between matter and radiation is what inspired cosmology (more specifically, the big bang model) to postulate that 15 billion years ago the universe went through a phase during which its properties were dictated by the electromagnetic radiation it contained, rather than by matter. During this so-called "radiation era," the structure and evolution of the universe were determined by the relativistic energy of the electromagnetic radiation in which it was bathed.

As we have just described, general relativity is, first and foremost, a theory of gravitation. But this interaction becomes one with the very structure of the space-time continuum, whose geometrical properties play a decisive role.[6] That is probably why Einstein wanted to go a step further by trying to geometrize matter so as to incorporate it in this synthesis. He long sought, but in the end failed, to find a unified theory in which matter itself would be reduced to geometry.

In any event, general relativity provided the key to numerous syntheses, including space and time fused into space-time continuum, gravitation and geometry, gravitation and light (since both propagate at the same velocity *c*), and (lastly) light, matter, and energy.

Mechanistic Conception and Its Failure

Descartes had clearly spelled out the mechanistic credo: "The universe is a machine in which there is nothing else to consider but the shapes and movements of its parts."[7] It is undeniable that Newtonian physics successfully achieved at least part of this program. In its wake, the unitary dream of the nineteenth century took on a decidedly mechanistic character. Helmholtz, for one, affirmed:

> We finally arrive at the discovery that the problem of the physical sciences consists in reducing natural phenomena to invariant forces of attraction and repulsion, the intensity of which depend exclusively on distance. The solution to this problem is the prerequisite for a complete understanding of nature.... The mission will be fulfilled the moment the reduction of natural phenomena to simple forces will have been completed and the proof established that this reduction is the only one consistent with the various phenomena.[8]

Naive as such an objective may seem, it did at the very least produce the kinetic theory of gases, which turned out to be an extremely powerful tool and was spectacularly confirmed by the observation of the Brownian motion. This success is definitely to be credited to a mechanistic worldview. Yet, pure mechanism has also met with resounding failures. Perhaps the most widely known is the one experienced by Descartes, whose theory of vortexes never proved adequate to account for a number of natural phenomena and was ultimately discarded in favor of Newton's ideas. The lesson to be learned is that a great unifying principle is not always enough for physics to move forward!

For many centuries, electricity, magnetism, and light resisted incorporation into a mechanistic vision. Newton himself opted for a corpuscular description of light. But despite a few early spectacular successes, his views did not prevail on this particular issue. Huygens had picked the right choice about light. He maintained that it was a wave "similar to those that form on the surface of water when one tosses a stone into it."

The mechanistic model proved wanting in several respects. Neither electrical or magnetic fluids, nor light particles, nor anything resembling the obscure ether were ever observed. The notion of spatially extended fields, such as described by Maxwell's equations, ended up winning the day for the purpose of describing these phenomena. There was no need to invoke any mysterious action at a distance or other exotic material substrate. Waves came to be interpreted simply as field variations. In such a picture, paradoxes disappeared, at least temporarily.

Descartes's ambition had been to extend the reach of the mechanistic approach to cover all disciplines. He boldly believed that all life-sustaining physiological functions in human beings (such as breathing, sleeping, feeling emotions, and so forth) "follow naturally in this machine entirely from the disposition of the organs—no more nor less than do the movements of a clock or other automaton, from the arrangement of its counterweights and wheels."[9] Following the success of Newton's mechanistic model, the dream was kept alive by an endless stream of ill-advised attempts to apply the same method to chemistry, biology, psychology, the social sciences, and economy.[10]

4

Modern Unifications

The Quantum Revolution

Quantum theory was invented to enable physicists to account for phenomena taking place on microscopic scales. These phenomena may involve either atoms or particles such as protons, neutrons, electrons, or even photons, which are discrete units of light. None of these objects can be seen with the naked eye, a magnifying glass, or even a microscope. They were not discovered until the twentieth century.

From about 1920 to the present, quantum physics has been the discipline of choice to advance virtually every area of science, from solid state physics to nuclear and particle physics. It came into existence after physicists began to realize early in this century that the traditional laws of physics were unable to explain the behavior of atoms. In particular, classical physics could not account for the stability of atoms, nor for the mechanism by which these atoms can emit light. Electrons too seemed to have peculiar properties. Although they were supposed to circle around the atomic nucleus somewhat like planets around the sun, they did not appear to have well-defined orbits. In most cases, it looked instead as though they were smeared out in space.

It became

increasingly clear that, to reconcile all this, a drastic overhaul of physics would be needed. What emerged from this realization was quantum physics. Among the famous physicists who contributed to its development are Albert Einstein, Paul Dirac, Erwin Schrödinger, and Niels Bohr.

Quantum physics has evolved into an extremely powerful tool. It has become the workhorse of particle physicists, nuclear and atomic physicists, solid state physicists, and even astrophysicists. None of its predictions—including some rather strange ones—have ever been contradicted by experiments so far. Because it relies on concepts that often have no equivalent in everyday experience, some of its predictions fly in the face of common sense.

We have already mentioned how the notion of particles had been losing ground to a harmonic description of physical phenomena. What we often refer to as particles—things like quarks, photons, and such—are really only fragmentary and limited aspects of more complex and profound entities.

Classical physics (as opposed to quantum physics) deals with two categories of objects—corpuscules and waves (in the classical sense). A corpuscule is a discrete object, localized in a very restricted portion of space, almost like a point. It follows a trajectory along which both its position and velocity are specified at any given instant. A wave, on the contrary—be it electromagnetic, acoustic, or any other kind—is not localized at all. It has a continuous character and occupies an extended region of space. In addition, two or more waves can combine by "superposition." The sum of two waves (of the same type, of course) has a perfectly clear physical meaning. The situation is quite different for particles. It makes no sense to add two billiard balls at the same location, and even less to claim that the result is another billiard ball.

There seems to be nothing whatsoever in common between a wave and a particle. Yet, as strange as it may be, the quantum revolution demanded that the two notions become intimately and irrevocably interwoven, which in the end amounts to denying either one of them individually.

To explain the need for such an unlikely alliance, one often cites the celebrated "Young's double-slit experiment." Without going into its details, we will stress the remarkable fact that no classical description in terms of either waves or particles alone can possibly encompass all its implications. From a particle point of view, electrons must pass through either one of two slits before striking a screen. Since a chunk of matter can only go through one slit at a time, whether the other one is open or closed should have no bearing on its trajectory, particularly if the distance between the slits is much larger than the dimension of an electron. In reality, experiments flatly contradict this intuition. It turns out that the state of *both* slits plays an absolutely crucial role in determining the spatial distribution of electrons striking the screen. It is as though, before traversing a particular slit, electrons can somehow "sense" from a distance the presence and state of the other slit and adjust their path accordingly. They appear to have a mysterious ability to "know" something about the configuration of the entire experimental setup. To make matters even worse, it is completely impossible to determine after the fact which of the two slits a particular electron went through.

Since an interpretation in terms of particles is unsatisfactory, some might be tempted to conclude that the experiment involves waves. And sure enough, the distribution of electrons impacting the screen is perfectly consistent with an interference phenomenon, which does suggest waves. Unfortunately, this interpretation is not entirely supported by experimental facts either, since electrons strike the screen as an ensemble of discrete localized impacts, as opposed to the continuous distribution expected of a wave.

From the point of view of classical physics, the contradiction appears irreconcilable. If electrons are viewed as particles, any interference phenomenon is a mystery and the distribution of hits on the screen cannot be explained. If, on the other hand, electrons are viewed as waves, it is impossible to attribute specific spatial trajectories to them, and the fact that the screen registers individual impacts makes no sense. Neither description is adequate. The nature of electrons is neither particle-like nor wave-like.

As it happens, physicists are faced with a frustrating situation. Whenever they ask a question involving the wave aspect of the phenomenon, the answer provided by experiments turns out to be consistent with the properties of waves. That is the case, for instance, when electrons are made to interfere or diffract through slits. And whenever they ask a question involving the particle aspect, the answer turns out to be consistent with the notion of particles, as in the case of the localized impacts of electrons on a fluorescent screen. But no experiment has ever shown an electron to behave like a wave and a particle *at the same time*. Could it be that the nature of the experimental apparatus is what determines the way phenomena manifest themselves? If so, what is the "true" nature of electrons when no measurements are attempted on them? Through the bias of an experiment that is really extraordinarily simple, physics is led to ask profound questions about whether reality is at all accessible to us.

At the cost of tossing aside classical conceptions, the formalism of quantum mechanics is perfectly capable of accounting for this particular experiment and many others like it. The concept of wave function (or of an associated "quantum field," to be discussed below) allows us to describe the physical state of a system, without insisting that the electron be either wave or particle. The term "wave function" is perhaps a bit unfortunate because it tends to conjure up the image of an ordinary wave, whereas, in reality, a wave function behaves quite differently. In particular, it can undergo a sudden "collapse" during the measurement of a physical quantity.

Quantum Complementarity

It is always useful to have two ideas: one to kill the other.
—Georges Braque

It has been claimed that the notion of quantum complementarity is rooted in the "correlation of opposites" advanced long ago by Heraclitus. Complementarity may fit right into the purest philosophical tradition, but that does not prevent it from being extremely difficult to grasp; indeed, the relevance of the concept has occasioned spirited debates. In fact, during the first half of the twentieth century, it became the central topic of virtually every polemic raging about the meaning of quantum mechanics, and it was an unending source of division among the founding fathers of the theory. Planck, Schrödinger, Einstein, and Louis de Broglie were opposed to it. Heisenberg, Pauli, Born, and Dirac accepted it, although not entirely without reservations.

The principle of complementarity states that physical reality is not fully described by the picture of a wave or of a particle. Two different and seemingly incompatible points of view—waves and particles—are necessary to account for certain phenomena. It is impossible to try to apply these classical concepts in the quantum realm without encountering this fundamental dichotomy.

Bohr argued that the meaning of any concept can be defined solely by means of concrete experiments. They—and they only—determine whether a particular concept is or is not useful for the purpose of describing results.

Isabelle Stengers wrote on this issue:

> It is the singularity of each science that is at stake when complementarity is asserted in the statement "there can be no answer without a question." In other words, each science must respond according to its own standards to the test of such pronouncements as: "No piece of knowledge can gain its emancipation from the question that gave it meaning in the first place." In turn, no question can gain its autonomy relative to the choice from which it proceeds; and no choice can avoid taking into account its selective

character or acknowledging what it prevents from taking center stage so it itself may be allowed to take center stage.[1]

In short, a concept is nothing more than an operational tool. Devoid of sense on an absolute level, it is functional only in the context of a particular experiment and not another. It cannot be exported from one situation to any other. In the double-slit experiment, for instance, the concept of wave applies so long as we do not try to specify the path followed by electrons, and so long as an interference pattern appears. The moment we insert any device whatsoever to monitor which slit the electrons go through, the concept of trajectory can then be defined, but the interference pattern vanishes. At that instant, the concept of wave loses its meaning and becomes useless.

Quantum objects become double-sided entities to which it seems we can successively apply contradictory propositions without in any way placing logic in jeopardy. The concepts of wave and particle are mutually exclusive. But they are both necessary to an observer intent on analyzing an experiment in terms of classical—that is to say, prequantum—physics. Only through such double language can we hope to grasp the totality of the information various experimental devices can provide about a quantum system.

Complementarity emerges as a kind of irreducible paradox linking a concept with its very negation; it is almost a contradiction in terms. When he proposed the concept, Bohr is said to have wanted to offer the basis for a new epistemology. Indeed, he spent much effort trying to apply it to other fields besides physics, notably to biology. Anyone can appreciate, for instance, that it is impossible to entirely isolate a biological system from its environment without bringing about its death. A choice must be made between studying a living system and killing it. Along a similar line, Bohr also proposed a mutually exclusive relation between the practical use of a word and any attempt to define it precisely. He talked of a de facto complementarity between love and justice: Since justice presupposes emotional indifference, we can never be just with those we love. Many other similar examples come to mind, such as introspection and emotion (introspection dissipates the emotion it tries to describe), affectivity and thought, and individual and species. Not surprisingly, such naive attempts to generalize drew almost immediate fire.

Is Quantum Theory Complete?

Einstein felt that quantum physics could not possibly be complete on the grounds that it fails to describe the reasons for the behavior of individual systems and does not go beyond predicting the properties of matter from a statistical perspective. He articulated his objections in a famous 1935 scientific article known as the "EPR paper," after the initials of Einstein and two of his collaborators at the time, Boris Podolsky and Nathan Rosen.[2] He devised a thought experiment whose purpose was to demonstrate that quantum theory does not tell us everything we should come to expect from a sound physical theory. He started with three premises:

1. *The predictions of quantum theory are correct.*
2. *No effect can travel faster than the speed of light.* This implies that, in at least some cases, two events cannot possibly influence each other. Specifically, when the two events are so far apart in space and so close in time that light cannot possibly travel from one to the other, they must be independent. This hypothesis is often referred to as Einstein's principle of separability or locality.
3. *If, without in any way disturbing a system, we can predict with certainty (i.e., with a probability equal to unity) the value of a physical quantity, then there exists an element of physical reality corresponding to this physical quantity.*

Einstein was in fact proposing a very general criterion for reality. His understanding of the word "real," which conforms to common sense, can be summarized as follows: If we can predict the result of a measurement of a particular physical property, and if the prediction turns out to be consistently correct, we are then perfectly justified in concluding that the property at issue is not a dream or an illusion. It follows that such a property must have a corresponding element of reality.

Einstein went on to argue that for every element of physical reality defined according to the criterion just stated, there has to be a corresponding quantity in the associated formalism, whether the quantity in question is measured or not. Then—and only then—can a physical theory said to be complete.

Einstein showed that applying this criterion to quantum theory leads to the celebrated "EPR paradox." More precisely, when the set of three starting hypotheses is applied to a particular thought experiment, the resulting properties of a pair of particles are inconsistent with the predictions of the quantum formalism. It follows that this formalism cannot be complete. Einstein concluded that a better description of physical reality, yet to be discovered, must exist.

Niels Bohr responded by taking issue with the second hypothesis. He insisted that it could not be accepted unambiguously, which invalidated Einstein's definition of reality. It is impossible, he argued, to isolate particles from any experimental apparatus designed to measure their properties. Velocity, for instance, should not be considered an intrinsic characteristic of a particle because that property is shared between the particle and the experimental hardware. All a theory can claim to describe are phenomena that include in their very definition the experimental conditions used to reveal them, rather than any supposedly objective reality. That is precisely what quantum theory does best, since it inherently includes all possible predictions of experimental outcomes. As such, it is complete from a predictive standpoint. Bohr further admonished against any reasoning about the reality of things.

The Bohr-Einstein controversy raised contentious philosophical questions about man's conception of the world and the role of physical theories. But from a strictly scientific point of view, Einstein felt strongly that the incompleteness of quantum physics could only mean that a "better" theory would be worked out at some point in the future. As we now know, this optimism was not borne out. Worse, it was dealt two severe blows in succession: First, a theoretical discovery by the physicist John Bell (1928–1990) in 1965, and second, a series of unexpected experimental results.

As mentioned above, Einstein's three hypotheses lead to a presumed incompleteness of quantum mechanics. If these hypotheses are correct, which Einstein was absolutely convinced of, the quantum formalism had to be either replaced or improved. Bell, who was as much in the "realist" camp as Einstein, at first shared the view that events happen in and of themselves, independently of the experimental context of their manifestation. He was, however, struck by a peculiar observation: Theories purporting to "complete" quantum physics by way of supplemental (so-called "hidden") parameters, such as were proposed by Louis de Broglie or David Bohm (1917–1992), seem to be able to satisfy hypotheses (1) and (3) only by violating hypothesis (2) (such theories are said to be *nonlocal*). Bell began to wonder if this reflected particular deficiencies of the new theories or if it was a general property of all hidden variable theories.

Much to his surprise, he was able to prove rigorously that any theory claiming to describe reality on the basis of (1) and (3) is automatically in conflict with (2). Bell came up with a theorem that spelled out specific restrictions on what predictions can be made of certain types of measurements. These restrictions have come to be known as Bell's inequalities. Moreover, Bell determined that certain predictions of quantum theory violate these inequalities. For instance, quantum physics predicts that photons emitted simultaneously by excited atoms are more strongly correlated than suggested by theories involving local hidden variables. In due time, ingenious experiments were devised to resolve the Bohr-Einstein controversy, with all its philosophical implications, in the laboratory (sadly, only after the two protagonists had died).

A first generation of experiments in 1970, involving correlated photons, gave ambiguous results, primarily because of exceedingly weak signals. In 1976, Edward Fry used lasers to excite atoms and obtained cleaner results that seemed to support quantum theory. Eventually, in the early 1980s, a team led by Alain Aspect at the Institute of Optics in Orsay, France, did a series of experiments that showed irrefutably that Bell's inequalities did not hold. Quantum mechanics was fully vindicated. There are indeed situations in which two photons that have interacted in the past constitute an inseparable (or "entangled") system even when large distances separate them (although that is not true of all photon pairs). Their behavior is global, which precludes any description in terms of individual particles with their own predetermined measurable properties.

Einstein's separability and his criterion for reality cannot both be correct. Einstein's interpretation of quantum physics does not stand. In particular, the available evidence squashes any hope of "improving" the theory by adding local hidden variables. Quantum nonseparability has been firmly borne out by experiments. That is not to say, however, that its epistemological status or its connections with the fundamental postulates of physics have been clarified.

The Different Interpretations of Quantum Physics

What can be said about a reality no one can observe? This question is really at the heart of all debates about the meaning of quantum physics. Answers have tended to cluster around two poles. One is *realism,* which portrays physics as capable of exploring reality and revealing to us what exists independently of ourselves. Those who, like Einstein, embraced that position believed that "the object gives itself to the subject," and they had no doubt that the mission of physics is to develop an authentic vision of objective reality. The other pole is *positivism,* which holds that correct predictions of what can be observed constitute the ultimate essence of science. Everything else (such as metaphysical inquiries about what is real) amounts to semantic games. In this doctrine, the word "reality" has no meaning in itself. Therefore, it is pointless to discuss the true nature of things. For a positivist, science boils down to its efficacy. This stance is not too different from the one adopted by some of the founding fathers of quantum physics who, rallying around Niels Bohr, proposed what came to be known as the "orthodox" or Copenhagen interpretation—a reference to the fact that Bohr was from Denmark. Torn between these two schools of thought, physicists had a number of options.

Some simply endorsed the point of view of the Copenhagen school. Accepting the premise that there is no such thing as inherent reality, they simply refused to engage in discussions on the issue, which (as far as they were concerned) were doomed from the start. Quantum physics works, and that is all that can be asked of it. There is no need to insist on any deeper cognitive implication.

Others grappled with what might be called constructive malaise. For them, the quandary created by quantum physics was proof that it can only be an approximate theory. Problems of measurements, collapse of wave functions, shattered determinism—all these difficulties were just too hard to swallow. The only solution was to revise the theory itself. But any alternative had to meet a tough challenge: It had to do at least as well as the existing formalism, which no one denied worked quite satisfactorily. The task became even more daunting after Bell proved that hidden local variables were not the answer. Any counterproposal had to include nonlocality, or nonseparability, if it was to be at all credible. One such example is David Bohm's theory, which is the archetype of so-called nonlocal hidden variable theories.[3] By introducing certain modifications to quantum concepts, it reinstates strict determinism. However, many physicists remain skeptical about the theory, as it raises a number of difficulties of its own.

Still others have proposed to reexamine the deeper implications of quantum physics. Some accepted the quantum formalism at face value but contested the Copenhagen interpretation, which they deemed too minimalist. Eugene Wigner (1902–1995), for instance, claimed in 1962 that the collapse of a wave function forces us to consider that our conscience may have a direct influence on physical reality. He asserted that the cognitive act of an observer endowed with a conscience is what causes the wave function to collapse. The problem with such a spiritualistic thesis is that it is far too vague. It lacks a sound theory of conscience, which seems to be nowhere in sight.

An even more esoteric interpretation was developed by a few other physicists. It is based on the notion of parallel universes. The first such theory was advanced in 1957 by Hugh Everett. It proposed that whenever a measurement can produce one of two results, the system made of the measuring device and the object measured would split, effectively creating two parallel universes. One of the two outcomes would materialize in the first universe, and the second in the other. In this picture, all possible results would be realized, at the cost of duplicating the universe. This theory may seem far-fetched, but it is as difficult to refute it as to confirm it. That it was conceived of for the specific purpose of trying to resolve the problem of measurements in quantum physics is a testimony to the bizarre enigmas posed by the theory.

The boundary line dividing those who accept the quantum formalism from those who reject it is not always clear-cut. There are those who have attempted to show that the collapse of a wave function, far from being an arcane event of mysterious origin, is a phenomenon that physics can account for. A number of physicists, including Wojciech Zureck, James Hartle, Roland Omnès, and Murray Gell-Mann, have proposed something called "decoherence theory." Its goal is to try to explain why macroscopic objects behave classically while atoms and other microscopic particles behave in accordance with quantum laws. The theory ascribes a special role to the "environment," meaning everything (from the air they travel through to the ambient radiation) that bathes these objects. In particular, interactions between macroscopic objects and the surrounding radiation would have caused them to rapidly lose their quantum properties, the environment acting somewhat like an observer. As such, a macroscopic object and its environment must be treated as a global system (in contrast to the case of a microscopic object), and this "decoherence" effect would be responsible for making their motion classical. As is the case with most theories of this type, this one also has its proponents and its detractors.

It is not our purpose here to render a verdict for or against any particular theory, all the more so since numerous recent experiments have uncovered a spate of new facts, which remain to be digested from a theoretical standpoint. We simply wish to promote a spirit of tolerance: Let the advocates of strict realism refrain from claiming as absolute dogma that what is real must be comprehensible in its totality; let the radical positivists avoid condemning any inclination to deal with reality as a contemptible metaphysical exercise.

Fermions and Bosons

Two macroscopic objects can never be perfectly identical. Two billiard balls of the same color will always have tiny differences (such as a scratch or an indentation), none of which will influence their properties or behavior. Put another way, they can always be distinguished from one another. Besides, their respective identities can be made even more obvious with individual markings. In any event, markings or not, the fate of any individual object can always be perfectly known by visually tracking its motion.

Things are quite different in the world of microscopic particles. (The term "particle," taken in its broad sense here, includes atoms, atomic nuclei, composite or elementary particles, and the like.) There are two reasons for this. The first is that particles become indistinguishable.

Replacing one electron by another in an atom does not in any way change the properties of that atom. There is nothing to differentiate one electron from any other. It is impossible to identify one specimen by painting it in a distinctive color. The second reason is that the very notion of trajectory becomes meaningless in quantum physics. It is therefore impossible to track a particular member in a crowd of identical particles and follow its individual path.

Quantum physics provides a partial answer to the problem of identical particles by segregating them into two categories—fermions, named after the Italian physicist Enrico Fermi (1901–1954), and bosons, named after the Indian physicist Satyendranath Bose (1894–1974). A particle must be one or the other; there is no other choice. All electrons, protons, and neutrons are fermions; all photons are bosons.

What are the differences between the two types? In the framework of quantum mechanics, any system can be represented by a mathematical object specifying its state. This object is called an eigenfunction. Consider a system made of two identical particles and described by its eigenfunction. What happens when the two particles are exchanged? If the particles are bosons, the eigenfunction remains strictly invariant; if they are fermions, the eigenfunction simply changes sign.

That difference may seem like a rather trivial detail, but it has extremely important ramifications because it is related to the spin of particles. (The spin is a property that translates into quantum language the ordinary concept of self-rotation.)[4] It has been shown that particles with half-integer spins (e.g., 1/2, 3/2, and so forth) are necessarily fermions, while those with integral spins (0, 1, 2, . . .) are always bosons.[5] Fermions and bosons, although seemingly separated by so little, turn out to have drastically different behaviors, particularly at low temperatures.

The Concept of Quantum Field

The concept of quantum field makes it possible to integrate certain elements of the theory of relativity, which paves the way for unification of that theory with quantum mechanics. Whereas we used to refer to things as waves or particles, matter or radiation, such notions have been replaced by various excited states of a "quantum field," which (according to the new quantum theory of fields) corresponds to the only physical reality. That is not to say that quantum fields are "true" entities: They really cannot be pictured other than through an appropriate formalism. They may be the fundamental objects of modern physics, but their properties are purely mathematical. They add up to operators whose eigenvalues correspond to the various possible outcomes of measurements.

Although they are indispensable to describe real phenomena, quantum fields do not evolve in ordinary space but, rather, in generalized abstract spaces. The technical jargon invented to talk about these ultimate entities of matter is ample proof that they have very little to do with those tiny balls that are all too often used to describe particles, out of habit or oversimplification. In truth, they cannot be represented by points or any other geometrical form. An individual electron is not localized at any one point in space; instead, its properties are described by a quantum field extended throughout space. The only situation in which an electron can be assigned a precise position is immediately following a particular type of measurement, as when its impact is detected on a fluorescent screen.

Most physicists do not even try to interpret quantum fields, even though they form the basis of modern physics. From a practical standpoint, and for specific situations, the image of a particle remains a workable approximation. It is possible to force physics to conform to a corpuscular description, as imperfect and restrictive as it may be. As we have seen, certain experiments reveal the presence of electrons at particular points in space. A common mistake is to believe that the electron whose localized presence suddenly materializes is the correct way to look at the reality one tries to explore. It is impossible to know ahead of time just where an electron will appear. All one can do is define a "probability" of its appearance, which can be calculated from the amplitude of the quantum field at the point of interest. But even that ostensibly straightforward step requires a number of conceptual gyrations. For instance, could it be that the observer himself, by the very fact that he is trying to detect an electron, causes it to appear—"creates" it, figuratively speaking—where it did not exist as such a moment before?

That does not stop scientists from routinely reasoning in terms of particles and their interactions, in other words, from using the simple language of classical physics. It is only through such a sleight of hand that one can describe interactions as transmitted by forces, as we ourselves will do shortly, and that one can try to unify them.

Strictly speaking, the quantum theory of fields would require us to renounce such language and prevent us from resorting to oversimplified mental pictures, conforming more closely to our intuition. Yet, the theory offers a far more synthesized view of the world. We have already pointed out how it reconciled the atomistic and harmonic pictures to such an extent that it might seem difficult to push synthesis to an even higher level. Moreover, it offers a coherent and unified description encompassing all particles. It allows us to do away with old distinctions, once considered fundamental, between particles and their interactions, or between matter and radiation.

Particles and Interactions

The theory of atoms proposed by the ancient Greeks (Leucippus and Democritus) has undergone the most extraordinary metamorphoses as it evolved over the centuries. It seems to have finally settled today as a mathematical construct that has almost nothing in common with its original version. The formalism with which particles and their interactions are described is called quantum field theory. Its two components are quantum physics and special relativity, both of which are completely at odds with ordinary experience.

Even though it is a fairly recent scientific discipline (it did not exist until the twentieth century), particle physics is based on a long atomistic tradition dating back to the Greek philosophers. According to their conception, which was reincarnated in many different forms, matter is composed of indivisible entities. Their etymological name is "atoms" (literally, "what cannot be dissociated"). They correspond more or less to the modern concept of elementary particle.[6] We have mentioned that contemporary physicists do not necessarily adhere rigorously to the spirit and letter of quantum field theory and find it convenient to continue using the concept of particle. Their goal is to categorize such particles and, hopefully, to uncover some links between them, thereby reducing the pool of "genuine" elementary constituents, that is to say, of independent elements. But by the 1960s, the field of particle physics was in the throes of chaos. Thanks to the construction of ever more powerful accelerators, experimenters kept discovering new particles (some of them highly unstable), which theorists did not know how to classify in a coherent structure and whose properties they did not understand in the larger picture.

Today, the landscape of particle physics is both broader and more serene. Two theoretical advances explain this shift. The first is the development of the theory of quarks.[7] It enables us to better understand the properties of a subset of particles governed by the so-called "strong" interaction. These are called hadrons.[8] The theory led to a dramatic reduction of the number of contenders to the title of elementary, from about one hundred to only six—six quarks, to be specific! The second advance concerns the electromagnetic and weak nuclear interactions. It was originally proposed theoretically, and ultimately proven experimentally in the 1980s, that these two interactions are in fact not independent. This discovery, hailed at the time as a crucial theoretical breakthrough, can be viewed as the first step toward the unification of all the forces of nature. It is sometimes put on a par with the contributions of Maxwell, who had shown that electricity and magnetism are but two aspects of a single electromagnetic interaction.

Physicists account for phenomena on a microscopic level in terms of interactions between particles (strictly speaking, as we have just seen, between their associated quantum fields). At the present time, they recognize four such interactions as fundamental. This classification scheme may appear arbitrary and without justification, but it has the enormous advantage that it works:[9] All known physical phenomena can be described in terms of gravitation and electromagnetism, both of which manifest themselves on macroscopic scales, as well as nuclear interactions operating on microscopic scales. Of the latter, the "strong" nuclear interaction binds the constituents of atomic nuclei, while the "weak" nuclear interaction controls certain radioactive decay processes. Before even attempting to give a unified description of these interactions, we need to understand the properties of each.

Gravitation

"Universal attraction" governs many aspects of our daily lives. It is responsible for such ordinary phenomena as our weight, the fall of material bodies, and the motion of planets. It is entirely negligible on the microscopic scales typical of particle physics. But its effect extends to infinitely large distances. It always manifests itself as attraction and is cumulative, meaning that it is proportional to the number of particles involved. This last point explains why it is exceedingly weak on the scale of a particle but becomes predominant on a human scale and overwhelming on astronomical and cosmological scales.

Yet, there are many physicists who believe that gravitation might also play a role on infinitesimal scales, smaller still than the dimensions of elementary particles. In particular, it is thought to have been a key player in the exceptionally strongly condensed conditions prevailing in the very early stages of the universe, a few moments after the big bang. Given the enormous density of matter under such extreme conditions, its intensity must have been comparable to, if not greater than, that of any of the other interactions, even on the tiniest spatial scales.

Unfortunately, at the present time, we do not know how to describe gravitational phenomena on microscopic scales because the effects of gravitation are complicated by quantum effects. So far, gravitation has resisted any effort to describe it in the framework of quantum physics, and there are good reasons to believe that this is not about to change. In fact, not only is there no unified treatment of gravitation and quantum phenomena, but the two theories are even in conflict in some respects. One can always choose to apply one or the other, but chances are that they will produce inconsistent predictions. This state of affairs is a strong incentive to come up with a synthesis of the two theories. But to date, no one has been able to devise a quantum theory of gravitation or, for that matter, even a theory of gravitation that would be compatible with the principles of quantum physics. Until further notice, gravitation and quantum physics remain at odds, and modern physics continues to rest on two distinct pillars.[10]

The Electromagnetic Interaction

Electromagnetic interactions result in either attraction or repulsion, depending on the sign of the charges involved. Although in principle they have an infinite range, in practice their effects cancel out at large distances—in contrast to gravitation—owing to the overall electrical neutrality of matter. They play a preeminent role on ordinary human scales. Among other things, they are responsible for the cohesion of atoms and molecules, and govern chemical reactions as well as the laws of optics.

Most particles are electrically charged or, like neutrons, have a small magnetic moment. They are, therefore, all subject to the laws of electromagnetism. This interaction is adequately described by quantum field theory, or, more precisely, by a subbranch of that discipline known as "quantum electrodynamics." Electromagnetic interactions are thought of as resulting from exchanges of virtual photons that are undetectable as real entities.

The Strong Nuclear Interaction

The stability of atomic nuclei is amazing. Each nucleon (proton or neutron) is held within a nucleus with such strength that its binding energy (that is, the energy required to knock it loose) reaches tens of MeV.[11] What is the source of such cohesion?

In addition, all protons in a nucleus carry a positive charge. The associated electrostatic forces would tend to push them apart and would be expected to result in the disintegration of the nucleus. What is it that offsets this powerful electrostatic repulsion? And what is the force that keeps neutrons, which have no electrical charge, so tightly entrapped inside a nucleus?

The force responsible for this nuclear cohesion, which neither electromagnetism nor gravitation can explain, has been dubbed the strong nuclear force. Its effects manifest themselves in an extraordinarily short time, of the order of 10^{-23} second. That is the time it takes for certain highly unstable particles, sometimes called "resonances," to decay into lighter components. There is no faster phenomenon known in nature.

The strong interaction is extremely intense at short distances, but its range is so limited—approximately one fermi—that it affects only those particles that are in the interior of a nucleus.[12] It is completely negligible on ordinary scales, which explains why it was not discovered until well into the twentieth century. Hadrons are the only particles subject to the strong nuclear force. All other particles, called leptons, are completely insensitive to it. To date, over 350 hadrons have been catalogued, almost all of them unstable.

In fact, some semblance of order and coherence has already been restored to this chaotic jumble of particles, thanks to the notion of quarks, considered the constitutive particles of hadrons. Much the same way as the almost infinite variety of chemical substances can be reduced to molecular associations of a hundred or so fundamental elements, hadrons are described as combinations of six elementary quarks. A rigorously correct description of those quarks has to be based on quantum field theory; as usual, depicting them as corpuscules is merely a picture of convenience (strictly speaking, quarks should not be thought of as tiny balls). Be that as it may, within the context of this theory, interactions between hadrons amount to a set of more fundamental interactions between their constitutive quarks.

Whereas electrical interactions are traceable to electrical charges, the strong interaction has its origin in what has come to be called "color charge."[13] Just as electromagnetic interactions take place by exchange of photons, the strong interaction is transmitted by particles dubbed gluons. The name was chosen to emphasize the fact that quarks remain strongly "glued" to one another when one tries to separate them. Confined to the interior of hadrons, they can never be isolated and observed individually, which is hardly consistent with the common notion of "elementary" particles. While individual quarks cannot be extracted in their free state, particles made of combinations of several quarks or antiquarks have been observed to be ejected from nuclei. Such quarks are said to be "hadronized."

The Weak Nuclear Interaction

Although it too has a short range, this interaction is quite different from the previous one. It is responsible for beta radioactivity, a process in which a neutron decays into a proton, an electron, and an antineutrino.[14] Since a proton is ordinarily part of a nucleus and remains trapped in it, the phenomenon most typically manifests itself in the form of an electron emitted by an atomic nucleus. The weak interaction plays a particularly important cosmic role since it initiates thermonuclear reactions that enable the sun—indeed, all stars—to produce the energy that sustains our life. The process can go on for billions of years precisely because the interaction is weak.

Its range is so extremely short—about one-thousandth of a fermi—that it is practically a contact interaction, a bit like a glue. But its characteristic time, the time required for its effect to manifest itself, is considerably longer than for the electromagnetic or strong nuclear interaction. That is why it is often masked by the others, which spring into action far more rapidly. The speediest process wins out.

The weak interaction is transmitted by massive particles (almost a hundred times heavier than protons) called W^+, W^-, and Z^0 intermediate vector bosons. They were discovered in 1983 at CERN.[15] The existence of these three particles had been predicted a few years earlier by a bold theory uniting the electromagnetic and weak nuclear interactions (in a manner similar to the unification of electricity and magnetism, which had cemented the reality of electromagnetic waves). In the next chapter, we shall describe the key features of this unification, which is all the more remarkable since electromagnetic and weak nuclear interactions are quite different at ordinary energies. The principles involved put on a firm footing the seemingly unnatural marriage of photons with zero mass (in electromagnetism) and intermediate vector bosons with a heavy mass (in the weak interaction).

The principles underlying this unification were fully borne out by experiments, at least in the range of energies accessible with the Large Electron-Proton (or LEP) accelerator at CERN. At an energy in the neighborhood of 100 GeV, electromagnetic and weak nuclear interactions turn out to have comparable effects. On the other hand, at much lower ordinary energies, the difference between the mass of photons and that of intermediate bosons becomes the determining factor, and the two interactions are then decidedly different.

The Unification of Forces

A multitude that cannot be reduced to unity amounts to confusion.
—Pascal, Pensées

To have been able to reduce the description of the known physical world to a set of four interactions is a magnificent achievement. These interactions are all that are needed to explain the almost boundless variety of molecules and atoms, themselves viewed as combinations of only three types of particles (protons, neutrons, and electrons). But theorists did not stop there. Spurred on by the belief that unity is the simplest expression of order, they dreamed of further unifying the description of the physical universe. In the twentieth century, the quest for unity was essentially focused on the unification of interactions. Gerald Holton commented that, in this respect, the main themes driving physicists to pursue unification are really not all that different from those that preoccupied the ancient Greek philosophers.[16] Perhaps the world of ideas is not as changeable as some believe.

Einstein, for one, wondered if "electromagnetic effects could be viewed as a geometric property of space-time." Such an idea had worked before for gravitation. And electromagnetism and gravitation are not entirely without similarities, if only because in both cases the law describing the relevant force is inversely proportional to the square of the distance. This inspired Theodor Kaluza and Oscar Klein to propose, in the early 1920s, a theory exploiting such similarities. They introduced a hyperspace with five, rather than four, dimensions, where electromagnetism stemmed from the curvature of the fifth dimension, in accordance with Maxwell's equations. To explain why we are able to perceive space-time in four dimensions only, they theorized that the fifth dimension folds back on itself on an extremely small scale (specifically, on Planck's scale, or 10^{-33} cm). Much as a piece of cloth—a legitimate three-dimensional object—appears two-dimensional because of the small diameter of its threads, space-time would have artificially shed one of its dimensions. As we will see shortly, similar ideas have recently made a comeback in a more elaborate form with the theories of strings and superstrings.

Other attempts at unification were proposed in the first half of the century. They did not meet with much success either. In 1946, Arthur Eddington published his *Fundamental Theory*, founded on the notion of a strong interconnection between microscopic and macroscopic phenomena. The theory is no longer taken seriously by either particle physicists or cosmologists. Paul Dirac had a continuing interest in such questions. Intrigued by coincidences between "large numbers," he proposed an ingenious theory that accounted for cosmological as well as microscopic phenomena. The theory failed to catch on because one of its key predictions—namely, that the gravitational constant G might change with time—was never experimentally confirmed. Nevertheless, Dirac's ideas remain among the most pertinent to these fascinating issues concerning hypothetical links between cosmology and fundamental physics. If they ever resurface at some future time, it will undoubtedly be in a form quite different from their original formulation.

At any rate, physicists continue to search tirelessly for a unified description of these four interactions. Having noticed that their respective intensities depend on the spatial scale or, equivalently, on the energy involved, most believe that there might exist a range of energies where these intensities should become comparable. That is the guiding principle behind current research. The goal is to devise a theory in which each of these interactions is one facet of a single generalized interaction.

Whether such a hope is a realistic prospect or a pure fantasy remains to be seen. But it has already cost considerable funds and efforts. The planned construction of huge accelerators will at least allow us to ascertain whether these are worthwhile avenues to pursue.

Gauge Theories

When Maxwell published his celebrated equations in 1864, he knew that he had just united electricity and magnetism into a common framework. That was in itself a great achievement. What he did not know was that the mathematical formalism he had developed would continue to guide physicists more than a century later. The reason is that Maxwell's formalism leads to the notion of gauge invariance, which has come to be a staple of modern theories.

Maxwell's equations involve mathematical constructs called potentials, which make it possible to calculate electric and magnetic fields as functions of position and time. These potentials are not directly observable quantities. For one thing, Maxwell's equations do not determine them absolutely; only differences in potential matter. One can arbitrarily change the reference of the potential, or its "zero-point" as it is often called, without in any way affecting the laws of physics. Such modifications of the potential are referred to as "gauge transformations." Maxwell's equations are said to be "gauge-invariant" because they are unaffected by such transformations.

We now consider the case of quantum electrodynamics, which is the quantum version of electromagnetism, that describes with amazing accuracy a variety of phenomena involving things like electrons, positrons (the antimatter counterpart of electrons), and photons. The theory is governed by equations requiring that a "zero-point" be defined, similarly to what happens when one tries to define the coordinates of a point in space. The mathematical formalism that particle physicists use to describe particles involves a quantum field that must be assigned an arbitrary phase. (Since a quantum field is a complex quantity, it is characterized by an amplitude and a phase, much like an ordinary wave.)

Just as geographers insist that the distance between two cities not depend on their choice of origin, physicists also want their theoretical calculations to be the same regardless of any arbitrary choice of phase at every point of space and time. Such is the essential idea behind gauge invariance. As it happens, when one tries to blindly treat the case of free electrons and positrons, the results of calculations turn out to be very much dependent on the phase. Something needs to be done to get rid of this troublesome dependence. The trick consists in including in the equations an additional term whose purpose is to locally compensate for, and globally cancel out, the undesirable effects of the arbitrary phase. This additional gauge field turns out to be the electromagnetic field itself. This result leads to an unexpected way of interpreting the concept of interaction: The field that had been added artificially, in a manner of speaking, to eliminate any arbitrariness from theoretical predictions emerges as precisely the same one that is responsible for the electromagnetic force coupling electrons and positrons. It further turns out that the force is transmitted by particles characteristic of the electromagnetic field, namely, photons, and that the mass of these photons has to be equal to zero.

This example illustrates how the details of a given interaction can flow directly from the particular gauge invariance adopted. Generalizing to any other interaction, the problem boils down to finding the appropriate gauge invariance. The same type of procedure has been successfully applied to a number of fundamental interactions, including the weak and strong nuclear forces. In all cases, one requires that the equations describing these interactions remain invariant under a particular type of transformation in space and time. As happened in the case of quantum electrodynamics, such restrictions determine not only the actual form of the relevant interaction, but also the particles that transmit it. That is how the W^+, W^-, and Z^0 bosons came to be postulated for the weak nuclear interaction, or the gluons for the strong nuclear interactions. In short, the principle of gauge invariance is an extremely powerful tool. During the 1970s, it enabled the spectacular unification of electromagnetic forces and weak nuclear forces under the umbrella of a common theoretical framework.

Gauge theories constitute the latest embodiment of the quest for harmony, which is the underlying theme of this book. They interpret interactions in terms of symmetries, along a line similar to the one first followed by Kepler. Gauge theories are geometrical in character, provided that the word is understood in its broadest sense. They operate in abstract spaces that involve structures considerably more complex than those of "ordinary" space. As the mathematician Gilles Châtelet observed, the formalism constitutes the triumph of the geometrization of physics, a process that was begun three centuries earlier.[17]

The Electroweak Unification

Until the 1950s, electromagnetism and weak nuclear interactions were considered two separate interactions, by all appearances radically distinct in their phenomenology. Indeed, their intensities and the ranges of energy where they apply are quite different. Each had its own satisfactory theory describing it.

In 1957, hopes began to arise that weak interactions could be expressed in a formalism similar to quantum electrodynamics. This implied that "messengers," tentatively called W particles, should exist as counterparts to photons. The possibility of describing weak interactions in the context of gauge theories generated much interest, since any theory based on symmetries is always apt to please physicists and appeal to their sense of aesthetics. Describing several interactions with a common formalism was attractive enough. But the physicists Sheldon Glashow, Abdus Salam, and Steven Weinberg hit upon the idea of going one step further by unifying electromagnetism and weak interactions. They quickly convinced themselves that, although a number of additional hypotheses were needed, the goal appeared reachable. When the dust settled, new particles had been introduced, including a third messenger, or "intermediate boson," dubbed Z^0, and a particle with very unusual properties called the Higgs boson.[18]

Experimental confirmations of this new electroweak theory came in stages. First in 1973, so-called "neutral currents," proof of the existence of Z^0 particles, were observed in the Gargamelle detector at CERN. Ten years later, the W and Z^0 bosons were detected by a team headed by Carlo Rubbia, who was to receive the Nobel Prize for that achievement.

Physicists continue to debate the significance of the electroweak theory. Is it a true unification, or simply a collage of distinct laws? Whatever the case, it does not seem to go much beyond the level of formalism. While they may be described by a common theory, electromagnetism and weak interactions remain quite dissimilar. Not to mention that the theory requires no fewer than eighteen independent parameters! In the minds of many an expert, that is a nightmare comparable to the epicycles proposed long ago to explain the motion of planets. To be sure, it is quite exciting to have a theory capable of describing electrons and neutrinos (or *d* and *u* quarks) as merely two different energy states of a same particle. But such convenience remains largely abstruse, and numerous physicists continue to consider these two particles, which have drastically different properties (notably their masses), truly distinct.

Colliders to Explore the Origin of Masses

The symmetries invoked by the electroweak theory permit us to lump electromagnetism and weak interactions together. In its most simplistic implementation, the theory requires the existence of four mediating bosons, all of which should have zero mass. While photons are indeed massless, it is obviously not true of the three intermediate bosons W^+, W^-, and Z^0, responsible for weak nuclear interactions), which are routinely produced in particle accelerators. As a matter of fact, all three turn out to be quite heavy (nearly one hundred times heavier than protons). To account for this discrepancy, it is necessary to substantially complicate the theory by adding a number of ad hoc ingredients.

Some physicists have argued in the last couple of decades that the fundamental symmetry postulated in the standard model was "spontaneously broken" by a mechanism occurring early in the history of the cosmos. This kind of original sin, as it were, would have destroyed the primordial symmetry. All that would be left behind would be shadowy vestiges of residual symmetry. The responsible hypothetical phenomenon is referred to as the "Higgs-Brout-Englert" mechanism, after the three physicists who first proposed it. It would account for the mass of the W and Z bosons.[19] Electromagnetism and weak nuclear interactions would have become decoupled as a direct result of the breaking of this primordial symmetry.

Such a complex mechanism remains largely a mystery to most physicists, and no one—not even those who proposed it—has ever offered any justification or convincing explanation. Yet, it accounts very nicely for what is observed in real life. It even was able to predict the masses of the intermediate bosons long before they were actually discovered. It implies the existence of an additional field associated with a neutral particle (the Higgs boson) that should be observable under certain conditions, but whose existence remains to be confirmed experimentally. According to the theory, its mass should be no larger than 1 TeV. That turns out to be within the range of energies reachable by the planned circular proton accelerator, known as LHC (for Large Hadron Collider), to be built at CERN.[20]

The Unity of Physics

Coherence is the virtue of imbeciles.
—Oscar Wilde

From its inception, science has always taken the mantle of universality. "Any subject is one," proclaimed Buffon, "and no matter how vast, it can be condensed in a single discourse." Science's craving for unity acts as a most efficient stimulus to expand our knowledge. Indeed, it has led to increasingly spectacular successes.

Yet, modern science seems to have exploded into a multitude of highly specialized areas and distinct disciplines that may at times be interconnected, but that by and large ignore one another. There appears to be an overwhelming trend toward a proliferation of distinct and autonomous "subdivisions." Researchers in different fields often experience great difficulties understanding each other. Science in general seems to follow a course inconsistent with integration, partitioning its structure while at the same claiming to unify its content.

The contradiction is most striking in physics. One might legitimately question whether this drive toward unity is an inherent part of the scientific process. Could it, instead, just reflect intermittent and accidental episodes, perhaps only anecdotal, with no more than passing historical interest?

Modern physics has a thousand faces—mechanics, particle physics, astrophysics, and so many more—each offering some modest degree of unification, but hardly enough to triumph into a global unity. We simply do not have a truly unified view of the world, one that paints an unambiguous picture of some overall scheme. Each discipline explores one restricted aspect of the world, without offering so much as a clue as to how its hypotheses, fundamental concepts, and results might relate to those in other fields. The price to pay for scientific rigor seems to be a myriad of disconnected perspectives. Does this mean defeat? Should one acknowledge that the search for unity in physics has been a failure?

The Unification of the Objects of Physics

Before the seventeenth century, physics amounted to a rather crude generic science of objects linked by not much more than a few vague and tentative analogies. Following the birth of mechanics, however, it quickly began to flash tantalizing hints of its unifying potential under the banner of mathematics. By the eighteenth century, physics gradually extended its reach beyond its traditional boundaries, as Descartes set out on his ambitious agenda to develop the sciences of life without any recourse to mathematics at all. In his encyclopedic classification of the sciences, d'Alembert had made a distinction between the "physico-mathematical sciences," which were highly rationalized, and what he called "particular sciences," which concerned themselves strictly with empirical phenomena and factual data.[1] In either case, man's intellect was to consider any object "in the simplest and most abstract manner possible,...assuming and accepting nothing in such an object but the properties that science itself can suppose in it."[2] For d'Alembert, abstracting an object could only make it more amenable to complete intelligibility. Michel Paty observed:

> The rational sciences, with their special capability, would expand far beyond what d'Alembert had in mind. Electricity, magnetism, physical optics, thermodynamics, even chemistry, would all eventually lay claim to that privileged status. Indeed, the "particular sciences" described by d'Alembert have long ceased to exist except outside the confines of physics in the modern sense. The various branches of the physical sciences all seem to share a strong mathematical underpinning, at least as far as their concepts and the general form of their laws are concerned.[3]

It is by now generally accepted that the objects considered by the many different disciplines are actually one and the same, at least in principle and for the purpose of gaining some higher level of knowledge; this unique object is what we call "matter." Each discipline may consider it from its own angle, but a complete picture of matter can be attained only by considering all angles together.

The Unification of Disciplines in Physics

The degree of unity in physics and its ultimate goal have been perceived quite differently at various times in history. In 1908, Max Planck considered that physics essentially boiled down to two main branches—mechanics and electrodynamics. Each of these branches had completed a "horizontal unification," to use Edoardo Amaldi's phrase.[4] Planck began to wonder about a possible link between them. The phenomenon of emission and absorption of electromagnetic radiation by matter strongly suggested that there might indeed be one. If so, explaining how atoms emit radiation might hinge on a further unification of physics. Such a unification was unlikely to result from a mere juxtaposition of the two existing branches. Nevertheless, Planck looked hard for a solution in the framework of what was known at the time, avoiding giving undue importance to the quantum of action he himself had introduced in 1900. In the end, the required unification would be achieved only through a complete reassessment of the theory and a profound modification of fundamental concepts.

This example illustrates how at some stage of horizontal unification, which proceeds by merging two branches of physics considered until then separate, one invariably confronts a deep fissure that can be overcome only with revolutionary new ideas. Only then can a new explicative unity solve the problem at hand. In the particular example described above, quantum physics would unite matter and radiation in a common description. That unification operated "vertically" as it were, in the sense that the totally new concepts it gave rise to would find applications in disciplines as diverse as solid state physics and elementary particle physics.

A Universal Endeavor?

The spirit of science will be pushed to its limits, and its claim to affirm theses of universal validity will be negated by the evidence of these very limits.

F. W. Nietzsche, Philosophy in the Tragic Age of the Greeks

It is, no doubt, gratifying to speak of the "universality of science," but it behooves us to be on guard against an overly enthusiastic interpretation of the phrase. There are no such things as fundamental truths or superpowerful ideas from which all discoveries issue. There is no central concept that would explain atoms and galaxies, genes and cells, and everything else in-between. Reasoning does not enjoy the luxury of a unified system of categories. The universality of science is based primarily on the existence of a community of scientists, on a particular style of work, and on a shared way to approach solutions to problems.

Michel Paty observed:

> The unity of physics is not just what would exist after completing the ultimate synthesis, enshrined in the elusive equation of a unified field which Einstein was striving for in his later years. Inasmuch as such a synthesis might someday be achieved, it would clearly be premature at the present time. But, perhaps more visibly, unity is also a dream and a movement, and in that sense it is already fulfilled as it participates in the process of building a store of knowledge in physics."[5]

We will follow in Paty's footsteps and distinguish two distinct areas of physics that are open to unification. The first deals with knowledge in general and involves an integrated worldview. On this level, it is clear that unification has only begun and still has a long way to go. The other level concerns the methods themselves, the modi operandi. Here, there is at least partial success, which confers on physics—if not on science as a whole—what unity is required to guarantee its own identity. One could argue that the various disciplines, while short of providing a common and unified understanding of things, devote themselves to comparable pursuits and apply similar methods, even if it is to study quite different systems.

We have already touched on the role of mathematics, experimentation, and observation; on the power of the notions of harmony and symmetry; and on the validation of experimental results. These common factors undoubtedly contain the seed of the unity of science, through which the scientific community finds its intellectual homogeneity. Scientists constitute a distinctive group, which Jürgen Habermas referred to as a "communicating community," existing chiefly through a spirit of cooperation. In such a community, exchanges are supposed to be free-flowing, even though they may at times be slow. The history of science—physics in particular—is replete with examples of fruitful cross-pollination of ideas developed in separate contexts. Increased communication is bound to accelerate this trend.

The unity of physics, or of science at large, starts with the shared traits of those who chose this particular career and agreed on a common protocol, language, and set of beliefs. Julien Benda shrewdly observed: "It is highly significant that, when nations want to showcase their individuality and sophistication so as to impress their neighbors, they always flaunt their poets and artists; rarely do they draw attention to their scientists or philosophers, knowing full well that these are a common bond between humans, rather than a source of rivalry."[6] Science is indeed often perceived as the paradigm of reason, transcending national boundaries and cultural differences, and immune to the whims of shifting fads. "Science," proclaimed Frédéric Joliot-Curie, "is a fundamental element uniting human thought all across the globe. In my opinion, no other human pursuit commands as much universal solidarity."[7]

Descartes was one of the first to reject the traditional belief that different branches of science have their individual specificity. He reaffirmed unconditionally their creative unity inasmuch as they are endeavors of the mind and reflect the unity of human thought itself: "Instead of being an ensemble of different kinds of science, each having its own special degree of abstraction and intelligibility, and its own mode of certitude, it is a single universal science, completely one, as is the science of God, who sees everything in its essence."[8] Max Planck expressed much the same feeling in the twentieth century. In his view, the wonder of science is to avail itself of highly relative and evolving human means in order to grasp realities that are absolute, universal, and invariant. Most of the time, our measurements, instruments, and experiments have very specific purposes. Yet, "starting from all these data, the goal is to zero in on the absolute, on what is universally true, on the constancy hidden within them."[9]

Consider the case of elementary particles. Advances in that area depend on knowledge borrowed from all other disciplines; among other reasons, the experimental hardware has become so complex that it requires theoretical and technological expertise from many different areas of physics. In that sense, it is fair to claim that all of physics is embedded in any one of its branches.

This admirable cooperation does not alter the fact that it is very difficult to articulate a universal description—one that avoids the pitfalls of historical tradition—of the methods and norms defining science. One reason is that, as Gaston Bachelard pointed out, "it is precisely by changing methods that science manages to become increasingly methodical." Which raises an interesting question: Does an evolving universality still qualify as universality?

The philosopher Paul Feyerabend (1924–1994) definitely did not think so. He drew some rather radical conclusions from his own analysis of modern methodologies, such as described by his peers Rudolf Carnap (1891–1970), Imre Lakatos (1922–1974), and Karl Popper (1902–1994). For Feyerabend, the necessity to take into account many different points of view by itself negates the very idea of a rule:

> It is clear, then, that the idea of a fixed method, or of a fixed theory of rationality, rests on too naive a view of man and his social surroundings. To those who look at the rich material provided by history, and who are not intent on impoverishing it in order to please their lower instincts, their craving for intellectual security in the form of clarity, precision, "objectivity," "truth," it will become clear that there is only *one* principle that can be defended under *all* circumstances and in *all* stages of human development. It is the principle: *Anything goes*."[10]

Reason, which supposedly presides over the progress of science, would only be an abstract monster in the image of mythological gods. Put another way, since there is no magic recipe for discovering truth, Feyerabend argued that the very idea of rational method is a fantasy. His chief argument is that the history of science consistently belies all the methodological maxims proposed by philosophers, and that the only one that could possibly be applied—refutation—is not nearly as powerful as Karl Popper professed in his theory of conjectures and refutations. Even the ideal of coherence seems to Feyerabend incompatible with the course of science.

Feyerabend is undoubtedly correct when he points out that scientists do not automatically discard a theory just because it leads to contradictions or anomalies. But he forgets that these same scientists never feel comfortable with inconsistent theories, and that they always try hard to resolve contradictions by either modifying their hypotheses or coming up with new ones. On a more positive side, Feyerabend's broadside is beneficial in that it might awaken those who tend to lapse into a dangerous dogmatic somnolence. But his criticism probably goes too far, inasmuch as abolishing norms would also eliminate the possibility of any critique: A critique only makes sense against the yardstick of a norm. What is more, the fact that any rule is debatable does not imply that all rules are equivalent. Astrology and astronomy have common roots, but no one would dare conclude that they belong on the same level. Speculating on the origins of science, Jean Ladrière noted: "Whereas it is true that science remains firmly founded on these originative forms of experience, it has nevertheless acquired its originality by systematically distancing itself from purely conjectural and interpretative modes of learning, and by developing its own methods to build knowledge."[11]

Mathematics
A Path toward Unification

If you were to seriously scrutinize matter, you would realize that, like a talented comedienne, all it does here on earth is play all kinds of roles, under all kinds of disguises.

—*Cyrano de Bergerac,* Voyages dans la Lune

The history of physics reveals a leitmotiv that may well be the primary factor of its unifying power. Galileo had understood early on that its backbone was a process of mathematization: "The book of nature is written in the language of mathematics...without which it would be impossible to understand so much as a single word." It would be a bit fastidious to review the innumerable examples of the preeminent role played by mathematics in the physical sciences, particularly in the modern era. It seems that every practitioner of physics has had to wonder at some point why mathematics and physics have come to be so closely entwined. Opinions vary on the answer.

"What is there for the human mind to grasp except numbers and measures?" Kepler asked that question almost as an admission of weakness. It is as though one does physics, particularly mathematical physics, because that is the best we can do. Bertrand Russell acknowledged as much: "Physics is mathematical not because we know so much about the physical world, but because we know so little: it is only its mathematical properties that we can discover."[12] Once again, it seems that the strength of physics stems from its wisdom to refrain from describing what it cannot describe. In Kepler's words, "geometry is immutable; it reflects divine intelligence. That men are able to participate in it is one of many signs that they were created in God's image." Later on, Kant's philosophy was to shed new light on this question, but no one has yet truly succeeded in explaining—and it is certainly not for lack of trying—the stunning success of mathematics in the physical sciences.

Mathematics may be indispensable to physics, but it obviously does not constitute physics, and there are many possible routes toward unification. One of the most popular, as we have already pointed out repeatedly, is based on the perennial search for harmony, most typically by invoking principles of symmetry.

The notion of symmetry seems to be universal. It is one of those arguments that do not require any additional explanation, since symmetry is often considered an explanation in itself. Things are what they are because, that way, "they are symmetrical." Nothing more need be said. Physicists just love symmetry, and it is all the more appealing to them when it applies to areas that at first appear unrelated, thereby establishing bridges connecting them.

Nowadays, we obviously no longer ascribe any special harmonic properties or explicative power to geometrical figures (such as circles, spheres, and polyhedra) or to music. Physicists have become more likely to turn to mathematics, a far better tool to express symmetries.

The Crucial Role of Symmetries

Pierre Curie was among the first to systematically study the symmetries of physical states. He stated in a 1894 paper: "When certain causes produce certain effects, the elements of symmetry of the causes must be reflected in the resultant effects. When certain effects display a particular asymmetry, such an asymmetry must originate in the causes that gave rise to these effects."[13] Physicists eventually began to analyze the symmetries of the laws of physics. These can be quite abstract, which is particularly true in the world of particle physics. Symmetries are directly related to the dynamical properties of the pertinent physical systems, that is to say, how such systems react to a force. The fascination symmetry holds for physicists is not just its ability to create pretty pictures; nor is it an exotic spice they use to spruce up their innate proclivity to coldly classify everything they come in contact with. The fact is that symmetry arguments have come to be the very essence of modern interpretations of physical laws.

This new way of looking at things is less than a century old. True, a number of physicists—such as Newton, Einstein, and Dirac—had succeeded in writing down equations describing the behavior of entire classes of phenomena. But there are fields, notably particle physics, where it is extremely difficult to wrest the applicable laws. For one thing, the pertinent forces are not always clearly known; for another, there are often limits to what experiments can be conducted. Under such conditions, identifying the symmetries governing the phenomena under study has proven to be quite a powerful approach. As we shall see, such symmetries are typically expressed in terms of "conservation laws."

Mathematicians have long studied geometrical symmetries, such as those displayed by crystals or the polyhedra considered by Kepler. They went on to discover numerous other kinds of symmetries, far more complex and abstract than those encountered in ordinary geometry. They were able to categorize them with the help of *group theory*, a branch of mathematics invented in 1830 by Evariste Galois (1811–1832). Most of these esoteric symmetries are difficult to visualize because they involve generalized hyperspaces constructed by mathematicians, which have very little in common with our familiar three-dimensional space.

Physicists were quick to adopt these abstract spaces and exploit their symmetries. They have learned how to describe the properties of particles in terms of specific characteristics of such hyperspaces. In particular, symmetry operations turn out to correspond to transformations of one type of particle into another, for instance, from proton to neutron. These two particles are obviously not identical, if only because of their different electrical charges. Yet, remarkable connections between them emerge from symmetry considerations. This implies that they share certain properties. They must be subject to the same type of interaction if they are to transform into each other under the right circumstances. The degree to which the latest physical theories rely on such strong links between the properties of particles—more specifically, the structure of the interactions that govern their behavior—and certain mathematical symmetries cannot be overemphasized.

Symmetries and Conservation Laws

A symmetry is defined as an invariance with respect to a certain type of transformation. A sphere, to take a simple example, can be rotated by any angle about any axis passing through its center without in any way changing its appearance. Such a rotation can be described mathematically by the equation of the sphere itself. Spherical symmetry results from the fact that the equation in question does not depend on angles of rotation and, therefore, remains invariant under any such operation.

This definition of symmetry leads to the all-important notion of *group*, in the mathematical sense of the word. A group is a set of elements that are transformed into each other via specific operations. Translations in space are a perfect example. Two successive translations are equivalent to another translation. The set of all transformations that leave an object invariant (i.e., that restore its original appearance) forms what is called the symmetry group of that object. Likewise, a physical phenomenon is said to obey a certain symmetry if its laws remain invariant under any of the transformations of the corresponding symmetry group.

A fundamental theorem, due to the German mathematician Emmy Noether (1882–1935), states that to any invariance in a particular symmetry group there must correspond a physical quantity that is conserved under the applicable transformations. In other words, the concept of symmetry goes hand in hand with conservation laws. If physical laws are assumed to be independent of time, for instance, the result of any experiment must remain the same no matter when measurements are taken. In the language of group theory, there is time invariance. Applying

Noether's theorem to this case leads directly to the principle of energy conservation.

Physical laws are further assumed to be the same in all locations of space. Translational invariance in space results in the principle of momentum conservation. Pushed a step further, this conservation law precludes any spontaneous change in motion, in keeping with the principle of inertia.

More generally, studying the symmetries of fundamental interactions (gravitational, electromagnetic, nuclear, and such) has led to fundamental advances in the last few decades. New concepts, typified by the notion of gauge symmetry discussed above, have broadened the range of applicability of group theory and have spawned revolutionary new ways of describing fundamental interactions.

P, C, and *T* Symmetries

Modern physics deals with symmetries that are in general considerably more abstract than the examples used in the preceding section. But for all their complexity, their role is no less fundamental. Cases in point are parity, charge conjugation, and time reversal.

We will start with parity. Consider a real experiment involving a collision between particles. Parity is an operation, denoted *P*, which creates—at least mentally—the mirror image of that experiment. In such an operation, the particles involved remain the same, but their positions do not since left becomes right, and vice versa. The question, then, is whether or not the mirror image of the original experiment can occur in nature or in the laboratory. Depending on the answer, the experiment is said to either obey or violate *P* symmetry. For a long time, physicists were convinced that all physical laws had to obey *P* symmetry. That belief was based on simple common sense, since nobody had any reason to doubt that the mirror image of any arrangement of objects could exist. Much to their surprise, they discovered in 1957 that the weak nuclear interaction in fact violates parity conservation, and nobody really understands why; it is the only interaction to do so. This violation makes it possible to define left and right in an absolute sense.

We turn next to charge conjugation. Every particle happens to have a matching antiparticle of similar mass but opposite electrical charge. Charge conjugation is that operation which transforms them into each other (e.g., electrons into positrons, or protons into antiprotons, and vice versa). This operation is denoted *C* to emphasize the fact that it involves charge reversal. Again, we start with a real experiment involving a collision between particles, and we apply to it the operation *C* according to the following recipe: Whenever we encounter a particle, we mentally replace it with its antiparticle counterpart, but we keep its original trajectory unchanged. For instance, a collision between a proton and a neutron is replaced by the very same process involving, this time, an antiproton and an antineutron. The original experiment is said to observe *C* symmetry if the new situation can occur in real life. The weak nuclear interaction, which we already know to violate parity, turns out to violate charge conservation as well.

Finally, time reversal, denoted *T*, corresponds to an inversion of motion rather than of time itself. It describes a phenomenon unfolding in a direction opposite to its normal chronology. Put another way, things look like a movie being rewound. The classical laws of physics tell us that if at any particular moment t_0 (taken as origin [$t_0 = 0$]) the speeds of every body in, say, the solar system (including the sun itself, the planets, and all their satellites) were suddenly reversed, the trajectory of each would remain unchanged, although its position at time $+t$ would, of course, become what it was at time $-t$ before the reversal.

Parity, charge conjugation, and time reversal all play a key role in the equations particle physicists routinely deal with, notably in the context of something called *CPT* invariance. As the initials imply, *CPT* is the product of the three individual operations *C*, *P*, and *T*, that is to say, a charge conjugation followed by parity, itself followed by time reversal. The fact that no law of physics is known to change under this operation is called *CPT* invariance. Although the weak interaction violates charge conjugation and parity separately, it nevertheless turns out to obey overall *CPT* symmetry.

In plain words, *CPT* invariance means that the known laws of physics also hold in a world made of antimatter observed through a mirror and in which time would run backwards. This property is fundamentally related to the principle of causality, which imposes some restrictions on the way a chain of events can unfold. Moreover, it implies a certain symmetry between matter and antimatter and predicts, among other things, that the lifetime of a particle must be precisely equal to that of its corresponding antiparticle. One question that has thus far eluded a satisfactory answer is why particles are apparently so much more abundant in nature than antiparticles. The "antimatter problem," which cosmologists have struggled with for decades, remains very much a mystery.

The Unifying Process

The relentless push toward unity manifests itself not just in science's operative method. Mathematization and harmonic considerations cannot explain it all. According to Pierre Duhem:

Since reason does not have access to nature's secrets, there is only one path open to the scientist: To conceive of intelligible connections between empirical laws; to deduce them from their logical independence or their partitionment; to make sure that all laws applicable to a given field form a coherent system. When such an approach succeeds, its benefits are twofold. First, it provides a more economical description of empirical laws; second, it suggests a logical (albeit artificial) classification scheme, inasmuch as these empirical laws can be elevated to the status of theorems or principles."[14]

In Duhem's opinion, if such a classification encompasses all known phenomena and proves capable of predicting new ones, it should not be deemed arbitrary or artificial. Instead, it qualifies as "natural." Finally, the logical structure of any physical theory suggests that it reflects a genuine or ontological order of things, although there is no reason to insist that the theory do so as a matter of principle.

The coherance of any discipline in physics ultimately manifests itself in the unity of its laws. The laws of classical mechanics, thermodynamics, quantum physics, etc., define each of these branches of physics and ensure their own unification. It sometimes happens, incidentally, that a particular law transcends the field for which it was originally devised. The law of conservation of energy, for instance, applies to all of physics.

What remains to be sorted out—and that task really falls to epistemologists—is how any particular unifying law takes shape in a physicist's mind. On a more fundamental level, who or what triggers the spark of understanding? Is there a transcendental mind that all knowledge can refer to as its source? Or do scientific discoveries come about through multiple agents? In the first case, all insights, no matter how diverse, would spring forth from a common seed reincarnated in many different individuals during the course of time. In the second case, the nature and content of individual contributions would matter more than the incidental personalities of those making the discoveries.

We are not about to propose answers here. We will simply stress the predominant role of imagination in the creative process: "Always on the alert, [it] can never stop organizing, at least in its broad outlines, the monogram of human understanding," declared Kant as he was formulating his theory of knowledge. The word 'imagination' does not mean here idle absent-mindedness waiting for truth to happen by, but an irrepressible impulse of the human mind. It is the powerful motivator behind inquiries about our world; it is the fire that ignites scientific thought itself. Imagination is entrusted with the responsibility to explore. Its mission is not to abjure reality but, rather, to magnify our intercourse with it. Karl Popper and Gaston Bachelard, among others, have emphasized the historical importance of speculation in forming hypotheses and establishing scientific truth.

The intriguing question is: What type of imagination is at play and how does it operate? Any method is legitimate to achieve useful connections, including fusion, juxtaposition, analogies, integration, synthesis, and whatever else might work. Even before the official birth of physics, powerful unifying currents inspired the then-prevailing worldviews. Mechanicism, animism, vitalism, harmony..., many such movements have led to important stages of unification in the history of ideas. Typically, attempts at unification have been sparked by analogies, whether consciously or not. At various times, the universe has been viewed as an army of gods, an organism, a machine, a clock, and more. For those steeped in physics, it could be a mechanism, a timepiece, or a computer. Even though such analogies are normally purged from scholarly discourse and official presentations, they often haunt our thoughts, sometimes suggesting connections apt to unite scattered knowledge. Pierre Thuillier cites the example of Lavoisier, who, having been appointed *fermier général*,[15] was used to dealing with revenues and expenses. Thuillier suggests that this ability to keep track of balance sheets might have ideally prepared Lavoisier to conceive of his famous law of conservation of matter.[16]

Analogies of this type sometimes amount to no more than caricatures and end up on the scrap heap. Yet, they have also sometimes helped develop important concepts—such as energy, information content, space-time, particles, and atoms—which proved essential to ensure the unity of a particular discipline or even to extend its reach to others. When it succeeds, a fresh vision is recognized as innovative and becomes part of the onward march of science. According to Thomas Kuhn (1922–1996), the hallmark of a scientific revolution (in other words, a discovery that makes science "progress") is a new paradigm, a new way of looking at things that makes it possible to explain results, which until then were inconsistent (or at least disconnected); it is a new unified vision, compelling enough to receive wide support and, in time, unanimous acceptance within the scientific community. That is one of the fundamental characteristics of any advance toward unification.

Enriching Syntheses

There is plenty of evidence in the history of physics that discoveries occur typically by way of unifying syntheses. The key to success often seems to be the fecund melding of two currents of thought considered independent, if not outrightly contradictory. We have already abundantly illustrated how the incompatibility between corpuscular and harmonic visions ultimately led to the advent of quantum physics.

Another productive antagonism revolved around the troubling issues of permanence and change. Heraclitus had promoted unification through the duality of being and becoming. To understand this union of opposites, Greek atomism offered a vision founded on *temporary* associations of atoms, which were *permanent*. This point of view has been fully vindicated by modern physics, which considers that any object or system is made up of elementary components that are immutable. Whether this is sufficient to account for all their possible properties—in other words, the whole issue of reductionism—is another question altogether. Much later, the philosopher Ernst Cassirer proposed an epistemological and philosophical approach purporting to "reconcile the two contradictory requirements of permanence and change."[17] It was to become the centerpiece of the type of unification allowed under his system, in which "unity and multiplicity are no longer antithetical entities, but two logical concepts requiring their synergistic existence."[18]

There is yet another connection that can explain, at least in part, Galileo's success. By simultaneously taking into account both theory and experiment through the bias of mathematics, he was able to break free of the stifling Aristotelian tradition, which had rested on experiment alone. Similarly, in the twentieth century, it was by establishing a link between mathematical models based on relativity theory and available experimental data that Georges Lemaître managed to push cosmology out of the rut it was mired in. The results of measurements of the red shift of galaxies, compiled by American astronomers, had been known for several decades. The relevant theoretical tool—relativity theory—had been published as early as 1917. Yet, it was not until just before 1930 that Lemaître put the two together and proposed his model of a universe in expansion.

Sometimes the association of ideas or facts that are contradictory, or at least without any apparent connection, induces an unexpected epiphany. A random jump of thought suddenly reveals as unified what only a moment before appeared multiple and disjointed. Kepler, Newton, Poincaré, and Einstein all testified to such happenings, recounting how an instantaneous insight unveiled to them a whole new way of looking at and connecting old phenomena, how they were driven to ignore the traditional course of action and rise above the paradoxes inevitably produced by too narrow a perspective.

The unity between different disciplines can also be evidenced by the broad validity of a particular model, or by the generality of a given approach. For example, no physicist would dispute the wide applicability of probability to physics in general. It guarantees the universality of the law of large numbers and legitimizes the fields of thermodynamics and statistical mechanics.[19] Another example is the so-called analytical approach, which made it possible to embrace in a common formalism disciplines as varied as Newtonian physics, electromagnetism, general relativity, and quantum physics. It even provided the means to bring to light the similarities of their deeper structures. Unfortunately—and that seems to be a universal phenomenon—what is gained in unity is generally lost in terms of intuitive vision of things.

There are many other phenomena that have acquired a validity extending beyond the confines of a single branch of physics. Examples include phase transitions and behavior at critical points, classification of singularities, equilibrium phenomena, scale invariances, renormalization processes, and hierarchies and interconnections of structures. Innumerable chemical species can be formed by molecules composed of no more than about one hundred atoms. Atoms themselves are composed of neutrons, protons, and electrons. Beyond this point, one descends into the depths of matter with the standard model and its quarks. Is there an even deeper level? Is there something to be gained by looking for it? Similar hierarchies, although perhaps less obvious, exist on astronomical scales as well: planets, stars, galaxies, groups of galaxies, supergroups, and perhaps other objects beyond that.

Unification sometimes presents itself in the form of a simple juxtaposition of separate disciplines applied to a common field. Such is the case of cosmology, which gathers in a grand summit specialties as diverse as "classical" cosmology, astrophysics, nuclear physics, and particle physics, to name but a few. It is not a true synthesis, as each discipline is called upon to contribute its own expertise without surrendering its specific identity. But the convergence of views resulting from this blending has enabled us to understand certain aspects of the evolution of the primordial universe; indeed, it gave birth to new concepts such as vacuum energy, symmetry-breaking, inflation, and cosmic strings. These concepts are almost like "monsters," in the sense that they do not fit into the mainstream of any discipline taken individually. Besides, we almost always lack a unified theory capable of correctly describing such concepts. We do not have, for example, a quantum theory applicable to curved space-time, which would justify the idea of cosmic inflation.

This artificial cooperation between different disciplines underscores the need for true unification. That would require researchers to not stop in midstream. On the topic of cosmic inflation, for instance, it is necessary to look for a new theory at the most fundamental level. Short of unifying gravitation with the other interactions, it should at the very least provide a common descriptive framework. Exercises in "tinkering" with cosmological models, which the innumerable existing variations of inflationary models add up to, have been, from this perspective, appallingly and hopelessly sterile. There are those who—intimidated by the admittedly daunting task of reconciling gravitation, cosmology, and quantum physics—prefer to invoke metaphysical or theological arguments (such as the "anthropic principle," "cosmic intention," or "divine intervention,") when dealing with issues of fundamental physics. It is everybody's prerogative, of course, to pick their explanation of choice, but the ensuing confusion thrust upon physics and its demoralizing effect are immensely regrettable: If I can "understand" a particular aspect of the universe on the basis of the "anthropic principle," why should I break my neck looking for a unified physical theory to account for it?

The telltale sign of a unification in depth is a more complete understanding of elementary objects and a wider reach to other fields and objects of physics. Nowhere was this more vividly illustrated than in atomic physics when matter became related to—indeed identified with—radiation. The science of atoms may be only one branch of physics, but virtually all other fields experienced spectacular advances the moment they could draw on the concept of atom and the atomic theory.

Chemistry, which had contributed much to the genesis of the atomic hypothesis, became closely intertwined with physics. The distinction between theoretical chemistry and physics became blurred. Yet chemistry has remained a distinct discipline with its own identity, largely because the complex and varied phenomena it deals with are generally not amenable to a simple quantum-theoretical treatment. For the most part, quantum physics is relegated to the status of hazy, albeit legitimate, guiding principle.

Likewise, condensed matter physics did not really take off until after the advent of quantum theory.[20] In rapid development since the end of World War II, it has become one of the main branches of physics, with a long list of practical applications that includes semiconductors, superconductors, magnetic memories, surface physics, phase transitions, and superfluids. Although it constantly refers itself to the atomic constitution of substances, condensed matter physics was able to develop its own specific concepts and methods because it typically considers entire systems as opposed to just their constituents. A good example is the concept of dislocation, which explains the plasticity of certain solids. It illustrates how two different structural levels can combine to give a new perspective: Dislocations are simply a macroscopic manifestation of the microscopic properties of atomic and molecular bonds.

Reality and Rationality

While pondering the type of knowledge imparted by physics, the German physicist Max Planck distinguished three different worlds: first, the world of "real" facts; second, the sensible world; third, the world of models and ideas. The latter is nothing less than the world of physics, whose function it is to develop as perfect a knowledge as possible of the other two worlds. The problem is that physics seems to be drifting increasingly away from the sensible world, to such an extent that the two may eventually become impossible to connect. According to Roland Omnès, the history of physics shows convincingly that "the inexorable inroads of the formal," at the expense of the visual, constitute the price to pay for unity in the description of phenomena.[21]

The atom of modern physics, to take a specific example, is no easy nut to crack. Not amenable to usual images and metaphors, it fundamentally eludes ordinary language and descriptions. Generally speaking, physical reality shows what it is made of only through exceedingly abstract theories, seemingly forever closed to sensible intuition and whose comprehension is less than transparent, to put it charitably! Incidentally, this would tend to prove that it is intelligence, and not intuition, that enables us to grasp reality, contrary to Henri Bergson's thesis.

Julien Benda has observed that "one ends up with an exercise increasingly removed from everyday images and experiences drawn from our sensible world. The paradigm of such an exercise can be found in certain concepts of modern physics, which consist of pure algebraic expressions lacking the slightest connection with any fathomable reality."[22] Steven Weinberg echoed a similar sentiment: "This picture represents a nearly complete triumph of the field over the particle view of matter: the fundamental entities are the quark and gluon fields, which do not correspond to any particles that can be observed even in principle, whereas the observed strongly interacting particles are not elementary at all, but are mere consequences of an underlying quantum field theory."[23] That did not prevent Benda from emphasizing how much he "honors the human mind for its ability to venture beyond the imagination."

We simply do not know what the ultimate objects of human knowledge are. The question has been much debated, particularly in the seventeenth and eighteenth centuries, in the context of cognitive philosophies. The empiricists answered that it is our sensations. The rationalists assured us that, on the contrary, it is what our intellect can conceive of (things like extent, substance, and idea). Kant tried to reconcile the two points of view by contending that perception without understanding is meaningless, and understanding without perception is empty. In other words, the object of knowledge is the perceptual world, but seen through the filter of our intellect.

Are we just thinking up a world that conforms to our mental structures, which keep evolving with time, without ever being able to define a deeper reality? Or are we capable of discovering the true reality of the world? If we take the latter position, a "realistic" one in the philosophical sense, the purpose of science is to explore and unveil a reality supposedly existing independently of ourselves. Those who subscribe to that view are convinced that the mission of physics is to develop an authentic vision of objective reality. It must reveal "the secrets of the Old One," to use Einstein's phrase. If the perceived world is not the real one, one may legitimately wonder where the real world is to be found. In what obscure corner of the universe should one look for it? Nor is it clear whether the supposed rationality—by means of which we would like to think we can apprehend the relevant laws—is honestly discovered or is simply manufactured by us. In short, is the world truly rational, or does its genuine scientific grasp presuppose its rationalization?

The mathematization of science promises the most magnificent discoveries, but at the same time, it raises the thorny question of the relationship between the "thing in itself" and the picture science portrays of it. Do ideas and logical relations, which are implanted somewhere in our minds, have the same ontological properties as objects perceived in the external world? This question arises naturally in the context of unity. Auguste Comte was of the view that human thought is one all across space and time: "Logical laws...are, by their very nature, essentially immutable and universal, not only in all times and places, but also for any and all subjects, regardless of whether they are what we call real or chimerical. In truth, they find their way even into our dreams."[24]

Gaston Bachelard took exception to this conception by arguing that if reason indeed creates science, it also creates itself in the process. In other words, it has to suffer the consequences of its own discoveries, and, as a result, it must remain prepared to modify and adapt its own logical principles and structures: "Any real progress in scientific thought demands a conversion. Advances in contemporary scientific thought have brought about transformations of the very principles of knowledge.... The simplest framework of the understanding cannot persist in its inflexibility if one is to measure the new destinies of science."[25] Rudolf Carnap, a leading logical positivist, put it this way: "The question of the unity of science is meant here as a problem of the logic of science, not of ontology. We do not ask: 'Is the world one?' 'Are all events fundamentally of one kind?'... In any case, when we ask whether there is a unity in science, we mean this as a question of logic, concerning the logical relationships between the terms and the laws of the various branches of science. Since it belongs to the logic of science, the question concerns scientists and logicians alike.[26]

To get to the bottom of this debate would require giving a precise definition to the words "world" and "real." That might take us far afield, with no guarantee that the question can ever be resolved. These very issues were discussed at length not long ago in hotly contested debates about the proper interpretation of quantum physics. We refer the interested reader to the highly pertinent recent monograph by Michel Bitbol, in which the author concludes that, in the final analysis, there is no real inconsistency between matured empiricism and critical realism.[27] In light of these comments and given the intricacies of the issues involved, we will not dwell any further on this controversy, especially since sidestepping it does not preclude practicing science.

Instead, we now return to the more prosaic topic of physics and physicists. It has become customary to distinguish two types of physicists—theorists and experimenters. Physics being itself subdivided in a vast array of specialties, such a division actually masks the existence of a great diversity of profiles. Indeed, knowing that someone "is a physicist" tells nothing about what he or she really does. There is another more fundamental distinction that could be made between physicists, one based on two radically opposed schools of thought. In contrast to realism (in the sense we have just defined), positivism holds that the very essence of science boils down to correctly predicting observations. Everything else adds up to a esthetic thrills, rhetorical games, and plain hot air. Ernst Mach dismissed the notions of space and time, which he considered "idle metaphysical entities."[28] He wrote: "Colors, sounds, spaces, and time constitute for us the ultimate elements."[29] This philosophy, which maintains that the word "reality" has no meaning in itself, refuses to join in ontological debates on the grounds that they would inevitably be purposeless. In the end, science would reduce to its operating efficacy and be devoid of any ontological implication.[30]

Amazingly, these two movements, whose dividing line is sometimes hazy, have coexisted peacefully. The reason is perhaps that, as Michel Bitbol argued, realism and empiricism are not as antithetical as many people think. At any rate, few physicists declare themselves fully on the side of realism or positivism, and most walk both sides of the fence. It is primarily when they talk about their own discipline and what it purports to describe that they come through as realists. They take the position that the objects considered by physics, such as atoms and particles, really do exist. They confidently maintain that theories account for the world as it truly is, composed of myriad particles with an objective existence of their own and interacting with one another. Yet, they tend to adopt the methods of positivism by utilizing only concepts with a proven track record. In short, they act very much like positivists in the actual practice of their profession. When they do calculations, their sole purpose is to come up with accurate predictions. At such times, ontological concerns about reality are, if not impossible, at least an unwelcome distraction, which would interfere with one's train of thought and detract from operational rigor. Indeed, not too many minds are capable of pondering the subtle interpretations of quantum physics at the same time as they are manipulating its equations.

Such questions, as fascinating as they are endless, are therefore largely kept outside the purview of everyday science and are left mostly to philosophers and epistemologists, people who have the time to take sides and justify their choices. Scientists step aside more or less willingly, at least in the course of their normal activities. One of the strengths of physics is precisely that it does not demand answers to these types of questions in order to make headway. From this point of view, quantum physics represents an extreme case. It works perfectly well, but no one knows exactly what it talks about, nor what kind of reality it describes! Science does not explain the world; it is no substitute for either philosophy or metaphysics. It would be pointless to expect that reflections on new scientific data are about to provide a definitive answer to the perennial problems addressed by philosophy. It is, furthermore, obvious that science is powerless to create and justify itself, although that takes nothing away from its effectiveness.

Interestingly, this dichotomy does not seem to generate undue psychological anxiety because each side steps to the fore usually at separate times. During the practical phase of his work, a physicist has to at least pretend to believe in the reality and rationality of the world and the objects he studies, leaving aside endless questions about concepts that belong in the sphere of interpretation and would only encumber his immediate pursuit. Of course, there is nothing to prevent him from engaging in parallel reflections about the foundations of his particular discipline. The time and importance he gives to each facet of the issue tell more about his individual mind-set than any job title printed on his business card. Efforts to gain a deeper level of understanding can run into serious difficulties. How is one to decide, for instance, whether it is molecules or atoms—or even the elementary particles that make them up—that are real, unless the only true reality is to be found exclusively at the level of macroscopic objects? Any option is legitimate, but does choosing any particular one not automatically exclude all the others? In the end, the only reality whose existence cannot be denied is that of an undifferentiated whole, which encompasses all these various objects and systems. This whole is the universe itself, and recognizing it as such is the prerequisite of any scientific endeavor.

Yet, while the reality of the universe has to be acknowledged, cosmologists are the only ones to tackle it head-on. Any other discipline in physics is forced to split it, perhaps artificially but unavoidably, into simpler distinct systems. As science was first gaining a foothold, it began to dissect things and handle them one small piece at a time. One may decide to work on the scale of molecules, or particles, or planets, or even galaxies. But the choice must be made ahead of any practical undertaking. Why not just work on a human scale, some will argue? Why not indeed, except that it would bring us right back to anthropocentrism, precisely what science has been trying to rid itself of as much as possible.

In practice, each discipline defines its own approach by clearly segregating what is of concern to itself from all the rest. When doing particle physics, it makes sense to systematically overlook that these same particles are also an integral part of planets and living beings, and to focus exclusively on the microscopic level. It is impossible to do quantum physics without pushing all epistemological criteria aside and believing wholeheartedly—if only temporarily—in the reality of wave functions and quantum fields, and in nothing else. One cannot do a good job at cosmology unless one accepts the reality of galaxies and the universe and disregards their structure on a human scale. One cannot do mechanics without reducing the properties of all systems to their masses and a few other limited physical quantities, such as velocity and energy. Likewise for thermodynamics, nuclear or atomic physics. Each discipline, indeed each subdiscipline, has its own domain of reality that denies, or at least deliberately ignores, that of the others: To an astrophysicist, the world is a gas of galaxies; to a relativity physicist, it is space-time; to a quantum physicist, it is an unlocalized wave function; to a particle physicist, it is a collection of particles endlessly interacting with one another.

The science historian Jacques Roger goes so far as to consider the fact that physics gave up insisting on reality as one of its defining characteristics. According to him, science took over when physicists "resigned themselves to rationalizing appearances"[31] rather than reality, that is, the true nature of things. After all, modern science was born the day Newton accepted the mathematical law of universal gravitation without any clue as to the real nature of the relevant force. That trend has grown more widespread ever since. Modern physics forces us to take a fresh look at the matter surrounding us and leads us to raise questions about the very reality of the objects it considers (see below). In the end, like any intellectual pursuit worthy of the name, it ends up tripping over the question of being. It is an age-old controversy at the heart of endless scholarly jousts, and a tricky one at that, if only because possible answers are loaded with complex implications: The word "real" conjures up the concept of existence, which implies that of being, which in turn connotes reality. It has all the earmarks of a vicious circle.

The role of theory has become so dominant that it is forcing physics to confront an Oedipal complex over its close relationship with mathematics. Even when deprived of natural guides and a priori categories, the human mind keeps on inventing. By virtue of its very structure, physics constantly reminds us that any comprehension of the world must resort to indirect strategies.

But by being ever more disconnected from our visual and intuitive awareness, our formalisms are becoming increasingly abstruse. This is a relatively new development. All through the nineteenth century, physicists were able to conduct their work without asking too many hard questions about the deep meaning of their discoveries. They would observe what they believed to be this "great machinery of the universe" and set out to record its workings as accurately as possible. Granted, the phenomena they were concerned with did not always resonate with our intuitive perceptions—we do not have an innate sense of magnetism or of the polarization of light—but for the most part, they seemed agreeable to a coherent description. Even highly mathematical notions (such as velocity, acceleration, and temperature) could still be handled by common sense without conflicting with natural appearances.

Quantum mechanics shattered this comfortable state of affairs. It is far less transparent than any theory had ever been before. The relationship between reality and knowledge lost the (sometimes misleading) attribute of certainty it had enjoyed since the beginning of the scientific era. The invisible wall quantum physics seems to have erected between its formalism and actual physical events developed cracks through which all kinds of nagging questions and arcane hypotheses came rushing through. As the philosopher Edmund Husserl commented, the spotlight trained by forced mathematization onto science is also a source of confusing shadows. When the notion of particle gives way to eigenstates and quantum fields, what possible connection is left between the world and its representation? In the final analysis, what is truly real? Is it the ontological baggage or the formalism itself? It is as though, having learned to manipulate objects, science has forgotten how to become one with them.

"Isolationism"
A Prerequisite for Physics

Except in cosmology, no holistic vision of the world—one that is global from the outset—can bear fruits. The way physics proceeds can be applied efficiently only to systems that, as a start, must be recognized as particular. In order to restore a bit of order to what would otherwise amount to confusing chaos, it is essential to differentiate things: A table is not a chair; an electron is not a photon.

The very first step in such an approach is to select a specific category of objects and extract it from its global context. For something to be recognized and grasped conceptually and practically, it must first be clearly demarcated by simplifying the real world. We shall call this initial mandatory step "isolationistic." Its role is absolutely crucial, since it defines the objects of concern to a particular branch of physics by setting them apart from the rest of nature. It literally amounts to slicing a piece of our environment to define objects *ad libitum*, henceforth to be separated and distinguished from all the others. Once again, Galileo was a pioneer in this approach, stating that "subjective" or "secondary" qualities were to be discarded to retain only those attributes that were pertinent to the objects or phenomena of interest.

Before articulating, one must delineate. From that point on, man spoke of a cup on a table and the table itself as two distinct objects, and he described the properties and behavior of one independently of those of the other. Likewise, he agreed that it was not necessary to worry about what may be happening on Sirius when describing the fall of an apple in an orchard here on earth. And he would be quite willing to accept that the acceleration experienced by an electron in an electric field depends solely on the magnitude of that field where that electron is located. Such "restrictions" are so fundamental that they are almost never explicitly spelled out in any scientific discourse. They go without saying. They are a necessary precondition if one is to study interactions and discover laws. If not for the existence of classes of objects and distinct phenomena extracted from the whole, it would be impossible to discern any pattern, trend, behavior, interaction, or anything else physics is capable of speaking about. Objects have to be isolatable, at least conceptually, if physics is to be able to recognize them. Physics is thus, by definition, the diametrical opposite of a "holistic" approach, which, on the contrary, is based on the preeminence of the whole over any of its parts and deals with the universal interconnectedness of all different realities.[32]

Can physics go on forever along this path? Perhaps not, as the road may be blocked by quantum nonseparability. As we discussed earlier, there are certain circumstances when it is no longer possible to define individual objects. And even if we could, they would instantaneously interact with the rest of the universe, for example, with the quantum fluctuations of vacuum, with the diffuse cosmic background radiation, or with the universal gravitational field. The reader will recall that nonseparability, which is built into the quantum formalism, was experimentally confirmed during the 1980s. We have explained how two photons that have interacted at some time in the past must be treated as an entangled whole, even if they have traveled far apart. In such a case, it is not permissible to consider them individually as distinct entities. The global character of their behavior precludes any explanation in terms of independent photons, each with its own well-defined physical properties. It is as though the two particles remain connected by an invisible link that depends neither on space nor on time, and they cannot be conceptually separated. Any action exerted on one has an instantaneous effect on the other, no matter how great the distance separating them.

Subsequent Reconstruction

If it were to stop in this isolationistic phase, the scientific method would remain sterile. The process of dissecting and isolating must be followed by a reconstruction. As soon as physics has pried reality out, it has to reassemble and unify it. It must identify among all the objects it has just isolated elements of similarity, making it possible to apprehend them through a single description. Once separated, these objects reveal more readily what eventually makes them answerable to a common approach.

In this unifying phase, science in general—and physics more particularly—teases out analogies, equivalences, and sometimes plain identities, which enable us to reconstruct the world in a more unified and potent picture. Only then can physicists "grasp" all objects of the same class as such and describe them by means of laws that acquire a universal character. Gravitation is a typical example. An artificial satellite and a natural one react similarly to gravity's influence. From the point of view of mechanics, the adjectives "artificial" and "natural" convey no pertinent information and can be ignored. In a similar vein, physics disregards most qualities associated with the sensory world.

Physicists proceed the same way when they deal with the microscopic world. By comparing the properties of various particles, they look for what they have in common. Particles that seemed distinct turn out to be identical. For instance, there is no difference between one electron and another. Replacing the electron orbiting around the proton in a hydrogen atom by another will not in any way change the physical properties of that atom (or of the world, for that matter).

The notion of indistinguishability is the bedrock of quantum mechanics. It reduces the entire universe to replications on a huge scale (by a factor in excess of 10^{80}) of a limited number of fundamental particles. These replicas are absolutely perfect copies, not just similar objects like cars coming off an assembly line. It is impossible to stick a label on one particular electron to distinguish it from all the others. The property that identical particles are indistinguishable is known as Pauli's principle. It enormously simplifies the task of particle physicists, who no longer have to worry about keeping track of electrons that are big or small, brand-new or worn out.

This two-step operation—separation followed by unification—seems to work well enough. Physicists split apart and reassemble things in a single dialectical process that makes the strength and scope of the scientific method. Indeed, it is this very approach that defines science. We could say virtually nothing about the world, if not for the practice of dissecting it into distinct objects, or at least separate systems. Nor could we say anything if these objects did not readily reveal their similarities and mutual interactions. Perhaps the world can become comprehensible only through this dual process of separation and unification. Einstein himself confessed to being awed by this "miracle."

Breakthroughs of Reductionism

Man sometimes pretends to be matter, nothing but matter, a thinking machine, a gelatin with desires; and the more he persists in this claim, with no other ammunition than the resources of reflection and reasoning, the more he proves the sovereignty of the mind, the only thing capable of conveying meaning. For negating thought is itself a thought.

—*Vladimir Jankélévitch,* Le Paradoxe de la morale (The paradox of morality)

As we have seen, the first task (often subconsciously) of the scientist is to isolate. He sections, dissects, and slices through the complexity of the world to try to understand it. In the process, he inevitably destroys any underlying unity since he ends up with a collection of disparate pieces. But the kind of unity that is being sacrificed is that of a barren holistic approach.[33] The second step of the process is a reorganization, a reconstructive surgery of sorts: putting things back together and coordinating individual pieces, each of which is understandable at its own level and has a unity of its own that can hopefully be generalized.

Since the end of the seventeenth century, the atomistic ontology of Newtonian mechanics has been perceived as the exemplar of the natural sciences. Starting from fundamental laws that describe the behavior of things strictly on a local and elementary level, global properties are reconstructed mathematically on the basis of local ones via a two-way correspondence. With this strategy, the structure of a system becomes understandable in terms of the properties of its constitutive elements. As such, this type of reductionism (sometimes called "vertical") consists in reducing the description of a particular type of objects or systems to that of simpler and more elementary ones. As Amaldi put it, "the vertical unification refers to relationships between the descriptions of the same phenomena provided by theories belonging in layers of knowledge of different depth."[34]

The tack adopted by physicists boils down to analyzing matter in terms of its "smallest" constituents, such as molecules, atoms, or particles. They extract the smallest possible number of fundamental laws to account for the behavior of these elementary units. Reductionism is the belief that all phenomena are embedded in the properties of their constitutive elements, hence describable by just a few laws. According to this view, the description of "large" objects can be distilled to that of more elementary and fundamental ones. Nuclear physics would derive from particle physics (with its quarks and gluons), biology from molecular chemistry, chemistry from physics, and so forth. Such an approach has a profoundly unifying character since it condenses several disciplines—together with their corresponding laws—into their most fundamental common denominator.

An oft-cited example is the connection between thermodynamics and mechanics. The former used to regard heat as a spontaneous movement of warm toward cold; the latter described how all physical bodies, whether in motion or at rest, exert an action on one another. Reductionism made it possible to unite these two theories. Once it became clear that matter is composed of particles (atoms or molecules) whose motions are governed by the laws of mechanics, it was but a small step to equate heat with the degree of agitation of particles. After that, thermodynamics became simply a particular case of mechanics, and heat transfer could be interpreted on the basis of the laws describing how motion is transmitted among atoms and particles. In that sense, thermodynamics was truly "reduced" to mechanics.

From the seventeenth century on, scientists had no qualms about embracing this type of reductionism (which actually appears to have left its imprint on physics from the very start). Both Descartes and Newton adhered to it, and later physicists were to be strongly guided by it. It is doubtful that physics could have evolved any other way.

Most historians of science see Robert Boyle as the one who laid the foundations of modern chemistry by substituting a mechanistic model for the conception that had been championed by Aristotle and Paracelsus. Even though he too upheld the corpuscular theory of the structure of matter, Pierre Louis Moreau de Maupertuis (1698–1759) believed that purely material physical forces could not account for the basic manifestations of life. Awareness cannot possibly spring forth from the unconscious: "We will never be able to explain the formation of any organized body solely by means of the physical properties of matter," he declared in his *Système de la nature*.[35] The debate about this issue goes on to this day.

For Karl Popper, reductionism is "the ultimate goal of any researcher," the end point of the pursuit of knowledge.[36] Einstein similarly defended the notion that the unification of science is necessarily linked to a fundamental form of reductionism. He wrote in 1918: "The supreme objective of any physicist is to arrive at elementary laws with which the cosmos can be reconstructed by pure deduction." Rarely has such an ambitious goal been advertised so matter-of-factly!

Reductionism seems to be an integral component of physics. To do science amounts to reducing complexity to simplicity—scientists themselves are to be blamed for turning this truth into a cliché. From a practical standpoint, scientists tend to view reductionism as an indispensable tool, one that is generally effective and reliable. Again according to Karl Popper, any respectable scientist has to be methodologically reductionistic and must strive to integrate "facts that one discipline might explain on the basis of a particular ontology with the ontology of another discipline." When couched in those terms, there can be little doubt that reductionism should be expected to yield "the most fantastic scientific discoveries." Before the advent of statistical physics, thermodynamics was little more than an elegant and coherent set of empirical rules, but it lacked any fundamental justification. Statistical physics was ultimately able to give meaning to the pertinent concepts and, more importantly, to relate them to those of other disciplines, thereby integrating them into a broader unified vision. One could also point to the examples of the molecular structure of matter, the atomic structure of molecules, and the corpuscular structure of atoms. Even when it is impossible to calculate everything at a given level from what is known on a lower level, our understanding of things remains firmly rooted in this type of reductionism, which, if nothing else, guides and spurs on all practical research.

Is this vision too optimistic, perhaps even naive? Or does reductionism constitute the essence of any scientific pursuit?[37] Does it deserve credit for the most impressive discoveries of modern physics? Is it an indispensable component of the so-called "hard sciences," the *sine qua non* of their existence? Its obvious effectiveness would seem to suggest as much. All these questions are being hotly debated. No doubt, the correct answer, as always, is to be found in compromise and moderation; but we can already anticipate the need to distinguish between pure operational efficacy and the point of view of philosophy and metaphysics.

The Obstacles to Constructionism

On the strength of the remarkable success of the newfangled quantum mechanics, Dirac felt emboldened to write in 1929: "The underlying physical laws necessary for the mathematical theory of a large part of physics and the whole of chemistry are thus completely known, and the difficulty is only that the exact application of these laws leads to equations much too complicated to be soluble."[38] Dirac was underscoring a fundamental limitation—one with practical repercussions—of reductionism. Any macroscopic system contains billions upon billions of particles, governed by as many coupled equations linking their individual motions. No computer, however powerful, could possibly solve the equations applicable to such systems in order to derive their macroscopic properties. A reduction on that scale is totally out of the question. By contrast, macroscopic laws determined empirically, starting from general principles applicable to larger scales, prove much more tractable. Even today, some seventy years after Dirac's pronouncement, we still do not know how to obtain exact solutions to Schrödinger's equation, except in a few particular cases such as atoms or ions with a single electron. Does this merely reflect the lack of power of our computers, or does it imply some more fundamental barrier?

During a scientific meeting devoted to the foundations of quantum mechanics, held in Tokyo in 1983, the physicist A. J. Leggett, a recognized expert in condensed matter physics, argued: "Even in those cases where there is a widespread belief that a 'higher-level' theory can be reduced to a 'lower-level one,' e.g., that solid-state physics can be reduced to atomic theory and electromagnetism, this is a complete illusion. For example, I would challenge anyone in this room to prove Ohm's law rigorously, for a real sample, from the laws of atomic theory and electromagnetism."[39]

It would be a daunting task indeed to try to derive Ohm's law from first principles in a quantum formalism. It is far easier to establish it empirically, using simple experiments involving electrical circuits. Yet, no physicist would doubt for an instant that Ohm's law is in fact rooted in electromagnetic properties on a microscopic scale—in other words, in a few simple fundamental laws. If it were not true, physics would have neither relevance nor meaning. But to know the appropriate fundamental equations does not automatically mean that one knows how to describe the behavior of the objects they govern. It is not the equations themselves, but their solutions (which cannot always be worked out) that provide a useful description of physical phenomena. Just as knowing the equation is not the same as knowing its solution, knowing the formalism is not equivalent to knowing the phenomenon. It is invariably extremely difficult, and often impossible, to derive the properties of a complex system from the equations that apply to it. From a practical standpoint, this sets limits to how far reductionism can be pushed.

There exist many other similar examples. No one knows how to exactly calculate the long-term evolution of planetary orbits in the solar system. This in no way invalidates Newtonian physics. In the same vein, chemistry cannot be reduced in practice to atomic and quantum physics.

But it would be untenable to assert that the feat is impossible as a matter of principle. Likewise, nuclear physics cannot be reduced to the theory of quarks because physicists simply do not know how to calculate the concrete properties of atomic nuclei in terms of the laws governing the behavior of quarks. Yet, if someone came along who were smart enough to prove that it is impossible on principle, it would probably mean that our ideas about quarks are inadequate or even downright false. Indeed, what is there to ask of a theory of quarks if not that it properly account for the behavior of nucleons and atomic nuclei?

P. Oppenheim and H. Putnam put it more bluntly: "It is not absurd to suppose that psychological laws may eventually be explained in terms of the behavior of individual neurons in the brain; that the behavior of individual cells—including neurons—may eventually be explained in terms of their biochemical constitution; and that the behavior of molecules—including the macro-molecules that make up living beings—may eventually be explained in terms of atomic physics. If this is achieved, then psychological laws will have, in *principle*, been reduced to laws of atomic physics, although it would nevertheless be hopelessly impractical to try to derive the behavior of a single human being directly from his constitution in terms of elementary particles."[40] Wise words indeed, which have been paraphrased in many different ways, such as "Mozart's sonata in G major will never be explained in terms of the morphology of his brain." And so, the unity of physics may turn out to be not much more than a working hypothesis. To forget it and fall into ontological reductionism would be to renounce any possible scientific grasp of reality.

For all its effectiveness—or impotence, as the case may be—it is primarily on a metaphysical level that reductionism takes on its full significance. More precisely, as Karl Popper insisted, scientific research needs to be buttressed by a metaphysical purpose. There is nothing to contradict the view that different realities can truly be deduced from one another. And in truth, if this were not the case, why bother doing physics at all? However difficult it may be to reduce the description—we are not even talking of an explanation—of one level of phenomena to a lower level, it should not discourage anyone from viewing macroscopic systems as assemblages of microscopic constituents subject to their own laws.

To sum it up, there is no fail-safe recipe that enables us, in practice, to reconstruct a higher level from a lower one, nor to derive its laws. Enormous practical difficulties invariably crop up, the fundamental laws being of no use when it comes to tackling the concrete problems of mainstream science. It is, for instance, impossible to rely on such laws to describe the pion-nucleon scattering problem at low energies. Jean-Marc Lévy-Leblond, noted:

> Particle physics was born out of a desire to understand nuclear physics, which itself came into being because of certain problems in atomic physics. It was long hoped that by opening a set of Russian Matryoshka dolls, the enigmatic expression on the face of one doll might be explained by the next one inside it, and that by ultimately working backwards all the answers would fall neatly in place. In fact, the successive levels in microphysics turned out to have all the trappings of an inflationary spiral, and no recursion formula could ever be established."[41]

It is impossible to do without independent empirical data, even though hints of overlap are beginning to turn up between nuclear and particle physics.[42] Concepts such as symmetry-breaking, group renormalization, or scale invariance cannot be readily transported from a fundamental level, where they were originally developed, to an operational one. Our descriptions of the physical world seem organized in almost independent layers, as if concepts changed their nature as they pile up on top of one another. Perhaps that is why particle physics, once the crown jewel of physics, has lost much of its glamour; it is no longer considered the queen of science. The existence of the top quark, or of any other heavy particle discovered at CERN, is of no consequence to condensed matter physicists, no matter how much intellectual fascination it might elicit in them.

Complex things cannot be entirely understood on the basis of simple ones, even though the former are nothing more than an aggregation of the latter. Knowledge is not really structured like a set of Russian nesting dolls after all. The fundamental differences between large and small, between simple and complex, demand a change in strategy as the number of degrees of freedom of the system under study increases. Switching from one level of description to the next higher one generally calls for new principles, concepts, and procedures. As a bonus, these may even allow us to better understand how things work. But contrary to what the opponents of reductionism believe, it does not imply that the nature of things is qualitatively different. The properties uncovered by means of this procedure are sometimes referred to as "emergent." But it would be a mistake to accord too much meaning to the term.

The error would be no less deplorable than the one committed in the opposite direction by constructionists. There are some who, rather than confess their failure to understand, prefer to blame some "emergent complexity" as soon as a problem defies analysis. Abdicating their responsibility, they find a convenient refuge in "glorifying" this unexpected complexity, which they declare intractable on principle, instead of simply acknowledging their own deficiencies or those of their conceptual tools. Things are complex or "complicated"—the two are often confused—only because of our inability to comprehend them.

Mathematicians are well aware, for instance, that the simplest equations can hide enormously complex behaviors that are virtually impossible to study. Arithmetic, one of the most immediately transparent disciplines, illustrates the point vividly when the most penetrating questions about numbers can be formulated with disarming simplicity. How many pages did it take to prove Fermat's last theorem, although the theorem itself can be stated in a couple of lines and can be understood by any schoolage child?

By its very nature, isolationism disregards many of the properties of a system. This works only to the extent that physics sets itself a clear objective. If the goal is to describe how earth moves around the sun, it serves no purpose to consider the coherent motion of its billions of atoms. Applying the precepts of isolationism, a better game plan is to treat earth as a "simple" object with a mass, a velocity, and a moment of inertia. One might go a step further and split the system into some of its major components if the objective is, for example, to understand tides. But it would be absurd to study the behavior of each atom individually. As Dominique Lecourt observed, "physics has always constructed the objects it concerns itself with. This construction, which takes place through the bias of a particular experimental apparatus, always translates into a disconnection of phenomena, which requires that a number of properties be purposely excluded as 'negligible.'"[43]

It can thus be said that the object of any science becomes clear only after the science has articulated, at a sufficiently mature stage of its development, its own principles, that is to say, a few general propositions that form its basis. In Michel Paty's opinion, identifying that object constitutes for a science a genuine labor of unification: "The principles that define the very object of a science rationally—which make its apprehension possible—emerge from a process of simplification and generalization which comes down to a formula for unification."[44] It follows that the question of the object or objects of any given science is intimately related to formulating a unitary vision.

The problems besetting constructionism should not be construed as a failure of reductionism, which has come to be the primary guiding principle of physics. Once again, Michel Paty has this to say:

> To equate the search for the fundamental with a reduction to the most elementary object—and only with that—would be to fall into the trap of a positivist hierarchization of science. A "simple" object does not "simply" engender a complex one, for their imbrication works both ways. Besides, the notions of elementary, simple, and

complex are themselves products of the mind. They do not exist per se and are no more "natural" than any other notion. They merely guide our reasoning and enable some degree of comprehension of its object. Yet they can never capture its full essence, which remains beyond any direct grasp."[45]

By first isolating objects and subsequently putting them back together, physics implicitly extrapolates the macrocosm from the microcosm. This strategy seems reasonable enough, even though it is often impossible to actually reconstruct the macroscopic world that way. But to dismiss the validity of the approach would be tantamount to denying any relevance to much of science. To renounce explaining atoms on the basis of quarks and other elementary particles would be to betray the very spirit of physics and deny its effectiveness. In fact, it might even make sense to define the domain of applicability of physics precisely as that where reductionism works. Whenever it fails, physics seems to stop dead in its tracks.

The Limitations of Reductionism

The detractors of reductionism are probably mistaken as far as physics is concerned. On the other hand, it is difficult to deny that they have plenty of arguments at their disposal as soon as one leaves the sphere of physics.

Few would dare claim that biology is nothing but chemistry, that the phenomena of life can be understood as ordinary chemical reactions. Attempts to reduce biology to cells and genes, or intelligence to neurons, are not new. But by all evidence, they have had little, if any, practical impact. Those who believe they can describe life by dissecting inanimate objects that are fundamentally part of its constitution expose themselves to charges of oversimplification.

Genes, molecules, atoms, and ions are indeed crucial ingredients of life. Yet, knowledge (no matter how in-depth) of these entities does not constitute knowledge of life itself, which continues to enjoy a kind of "extra-territoriality," as François Dagognet put it.[46] No other object of physics possesses as much individuality relative to the others as does the realm of the living; as such, there is a great risk that physicalism pushed to the extreme might indiscriminately excise some of the properties of life. A mechanistic approach, which would isolate the substrate and give it a special status, seems ill equipped to deal with the animate world. But what is the best way to study life by itself, independently of matter? Paul Ricoeur made the following point, quoted by Jacques Testart: "From the

Renaissance onwards, only the non-living has been considered within the reach of our knowledge; the living must therefore be reduced to it; in that sense, all our thought processes are presently under the domination of death."[47]

To insist on understanding man in terms of cells alone, cells in terms of molecules alone, and so on does not seem a sure bet. In the meantime, it matters little to biologists that the protons, molecules, and cells they study contain quarks and gluons. They have to forge concepts that have no equivalents or counterparts in physics. Biology is not applied physics. Even from a metaphysical point of view, it is difficult not to see in life—and a fortiori in intelligence—the participation of something grander than what is called, significantly enough, "inanimate" matter. Reductionism is a risky proposition when applied to life sciences. Physics seems to be a far more natural habitat for it.

Quantum mechanics might also stand in the way of reductionism, but for different reasons. The gist of that approach is to reduce everything to microscopic interactions belonging in the quantum world. The evolution of a quantized system is described by wave functions and quantum fields. As we have seen, such objects cannot always be isolated. The wave function of a molecule cannot be reduced to a superposition of wave functions associated with the localized atoms or elementary particles that make it up, by virtue of the principle of nonseparability discussed earlier. Strictly speaking, it is not even possible to reduce molecules to atoms, nor atoms to particles, since in reality neither molecules nor atoms nor particles exist as such, but only under the guise of quantum fields! In this case, the failure of reductionism results from the inadequacies of our concepts. Nevertheless, as its very name suggests, quantum nonseparability may be fundamentally incompatible with the dissection of physical systems into simpler elements.

The Glamour of the One

That said, physics remains multiple and fragmented, a far cry from universal. Is this imperfect state of affairs temporary? The unity perceived by many physicists seems satisfactory enough, as numerous branches of physics appear more or less unified. Still, one may wonder if there might exist as yet undiscovered concepts and principles that could lead to a broader unification. One reason why the prospect for such a unification is far from certain has to do with the history of science. Most discoveries can be construed as syntheses. Yet, they almost invariably signal the emergence of new theoretical ideas and the prediction or observation of new phenomena.

Henri Poincaré observed at the turn of the century:

> Two opposite tendencies may be distinguished in the history of the development of physics. On the one hand, new relations are continually being discovered between objects which seemed destined to remain forever unconnected; scattered facts cease to be strangers to each other and tend to be marshalled into an imposing synthesis. The march of science is towards unity and simplicity. On the other hand, new phenomena are continually being revealed; it will be long before they can be assigned their place—sometimes it may happen that to find them a place a corner of the edifice must be demolished. In the same way, we are continually perceiving details ever more varied in the phenomena we know, where our crude senses used to be unable to detect any lack of unity. What we thought to be simple becomes complex, and the march of science seems to be towards diversity and complication"[48]

As we expand our knowledge and lump together in one global vision what were once distinct phenomena, new findings emerge without immediately fitting into this global picture.

We can thus begin to understand what is going on. Each successful phase in the evolution of science constitutes a partial victory for unification. At the same time, though, it unveils new elements that require a new and broader unification, so that the process seems endless. We might add that things are becoming more and more formalized and complex. What is lacking in the unity of the world cannot be blamed on the world itself, but rather on the inadequacies of our cumbersome thought processes. Luckily, since the notion of unity exists only in our minds, physics can continue to prosper and move forward.

Physicists have never forsworn their dream to give coherence to our view of the world. Even as scientific disciplines appear to multiply, the desire to unify the store of human knowledge continues to thrive. Erwin Schrödinger wrote in 1944, in the Preface to one of his books:

> We feel clearly that we are only now beginning to acquire reliable material for welding together the sum total of all that is known into a whole; but, on the other hand, it has become next to impossible for a single mind fully to command more than a small specialized portion of it. I can see no other escape from this dilemma (lest our true aim be lost for ever) than that some of us should venture to embark on a synthesis of facts and theories, albeit with second-hand and incomplete knowledge of some of them—and at the risk of making fools of ourselves."[49]

This raises a broader question: To what extent does such a goal remain within the purview of science, given that it reflects a motive that is primarily of a metaphysical order? Can a rational search for unity stand its ground against ontological affirmations of unity in this world?

Now more than ever, omniscience is either fantasy or hubris. We have no single system of illuminating concepts up our sleeves to give structure to the entire field of possible experience. Up until the nineteenth century, universities had managed to preserve the illusion of unified knowledge. Today's scientific landscape hardly conveys such a sense of cohesion. Some areas of physics, based as they are on apparently incompatible models, seem to have exhausted any prospect for further reduction. By all accounts, the world fashioned by physicists, ranging from quarks to masses of galaxies, has exploded out of control.

Yet, what would the stuff of science be if it did not single-mindedly pursue unifying foundations? Is there even a single physicist who does not look for a link that would enable him or her to think of two or more laws in a common framework? To reduce the arbitrariness of a description is to prevent the proliferation of causes.

While working on his theory of "holes" toward the end of the 1920s, Dirac initially thought they were protons, rather than electrons with a positive electrical charge. His motivation was that, until 1930, protons and electrons were the only two charged elementary particles known. For reasons of simplicity as much as aesthetics, Dirac was loath to introduce a new entity nobody had ever observed. He believed that if the proton could be assimilated with a state of negative energy left vacant by an electron, the number of fundamental particles could be reduced to just one—the electron. Such simplicity would have been "the dream of philosophers," he was to declare later on.[50] One never willingly chooses multiplicity; one has to be resigned to it. As it turned out, the correct choice was the positron, and Dirac ended up doubling the number of fundamental particles, ushering in the world of antimatter.

The existence of different theories for distinct areas of physics should not come as a shock, even if they deal with the same objects looked at from different angles. But a feeling of crisis might take over upon realizing how ambiguous, if not impossible, the smooth transition from one theory to another can be. To some, this lack of theoretical fluidity smacks of chaos.

There are those who consider microphysics more fundamental than any other area. They struggle with frustrating difficulties while they try to deduce from it the laws applicable on a macroscopic scale. Despite appearances of homogeneity because of a generalized use of mathematics, is physics not, in truth, splintered into heterogeneous specialties?

Despite recent declarations of victory, unity is not yet in the bag. Is it even reasonable to hope for a global formulation of knowledge? The very notion of an explanation of the world may seem utterly presumptuous. The history of human thought has been more akin to a series of stomach rumblings, interspersed with a few strident shrills, than a graceful and harmonious process of synthesis.

Dreams of Unity at the Close of the Twentieth Century

There is yet another evil I have observed under the sun; it is perhaps the greatest evil of all; it is the presumption of the human mind to want to explain the entire universe in four words, to capture the blue of the sky in a lekythos, and to force the infinite to fit within three fingers.
—Book of Ecclesiastes

The Utopia of the Whole

As we have mentioned, doctrines of things taken as a whole found a natural fit in monistic philosophies. Among them, we must make a distinction between those that hold that totality just is and those that promote unity through the bias of a transcendent principle. The universe is one by essence, believed a number of pre-Socratic philosophers. Those who agree with them that totality derives its unity and existence from itself will be inclined to make light of what internal multiplicity it may contain. Extremists might even argue that such multiplicity can only be rationalized as an incidental and temporary attribute. That, however, is not the point of view of modern physics, which considers that the unity of the universe stems from the laws that govern it. The physical order physicists talk about these days rings with a jurisdictive overtone: "Nature does not break its own laws," exclaimed Leonardo da Vinci, adding: "It is constrained by the rationality of its law, which is infused and lives in it."

At the end of the nineteenth century, Lord Kelvin, a distinguished scientist, professed to feel sorry for future physicists. Physics having achieved all its goals and reached its destination, he believed, there would be nothing interesting left for them to study. The turn of the century proved him grievously wrong. The advent of relativity, followed by that of quantum mechanics, put his naive and hasty judgment to shame. He should have learned from prior blunders. Francis Bacon had predicted as far back as the end of the seventeenth century: "There are in truth but a handful of phenomena specific to the arts and sciences. Discovering all causes and all sciences will be a task of only a few years."

Much later, toward the end of the 1920s, Max Born, eminent scientist in his own right, similarly assured anybody willing to listen that "physics will be done in six months." That forecast too was a bit off the mark. We now stand on the threshold of a new millennium and, once again, there are those who predict the imminent demise of theoretical physics. Since the grand unification of interactions is—or so they claim—on the verge of being completed, theorists will soon have no choice but to pack it in and join the unemployment line. Before long, the universe will have delivered its innermost secrets, and its majestic laws will thrive happily in a glorious ecumenical theory embracing the whole of the world under its benevolent mathematical gaze. We will finally know what the universe is made of, how it works, and where it is going.

Such proclamations have inspired recurring visions of an "equation of the universe," which the general public could display on T-shirts. The most dedicated believers might even have it tattooed indelibly on their chest. Granted, spectacular progress has been made in the last few decades toward a unified theory of interactions. But to trumpet the impending birth of a so-called *Theory of Everything*, with obligatory capital letters, may be going too far.

What really lies behind such pronouncements? Is the hope of reaching an all-encompassing view of totality not a bit brash? Does any discussion of the world in its entirety even belong in the realm of science? Since no experiment has been conducted, indeed can never be conducted, on the world as a whole, it is highly unlikely that the answers will come from experiments. The much-ballyhooed theory of everything may be condemned to forever remain pure conjecture. What facts could possibly challenge it? Should such a theory be put in place, it still could not hand us omniscience on a silver platter. In all likelihood, its equations would be so complicated that only in ideally simplified cases could they be solved. Therefore, if it is to be discussed at all, it should be with a sense of humility difficult to reconcile with its pompous label or with the scientistic demeanor of its proponents. Henri Poincaré had already expressed his own skepticism with an argument of a topological nature: "The scientist's brain, which is only a corner of the universe, will never be able to contain the whole universe."[1]

But the trap into which Lord Kelvin fell could be reactivated in the blink of an eye. The pressure to come up with the definitive answer is powerful and relentless, almost like a virus against which there exists no known vaccine. Yet, the list of problems facing physicists is lengthy, so lengthy in fact that any progress toward the ultimate revelation is at risk of amounting to something akin to a false pregnancy. Science is an ongoing process that, by its very nature, is incompatible with a static or definitive vision of the world.

Closure of what is endless is a contradiction in terms. Science is by essence a never-ending discourse, which is not to say that it does not experience time-outs. Even if by all indications physics had run its course, it would not have the means to come to that conclusion on its own. To pause is not the same as to have arrived. We would be well advised to mistrust the cult of the supreme goal and remain on guard against undue presumption. A single unexpected development or a new experimental result may well uncover flaws in what physicists had previously considered perfect. That would be enough to get science restarted on a brand-new march.[2]

The excesses of reductionism have abundantly demonstrated how easy it is to disfigure reality in one's zeal to force it into a unified mold. Even if unity were someday achieved, that would not necessarily shield us from the wrath Nietzsche expressed so eloquently in *The Gay Science*, as he lashed out at those who claim to explain all things solely on the basis of physical laws:

> It is no different with the faith with which so many materialistic natural scientists rest content nowadays, the faith in a world that is supposed to have its equivalent and its measure in human thought and human valuations—a "world of truth" that can be mastered completely and forever with the aid of our square little reason. What? Do we really want to permit existence to be degraded for us like this—reduced to a mere exercise for a calculator and an indoor diversion for mathematicians? Above all, one should not wish to divest existence of its *rich ambiguity*: that is a dictate of good taste, gentlemen, the taste of reverence for everything that lies beyond your horizon. That the only justifiable interpretation of the world would be one in which *you* are justified because one can continue to work and do research scientifically in *your* sense (you really mean mechanistically?)—an interpretation that permits counting, calculating, weighing, seeing, and touching, and nothing more—that is a crudity and naiveté, assuming that it is not a mental illness, an idiocy. Would it not be rather probable that, conversely, precisely the most superficial and external aspect of existence—what is most apparent, its skin and sensualization—would be grasped first, and might even be the only thing that

> allowed itself to be grasped? A "scientific" interpretation of the world, as you understand it, might therefore still be one of the most stupid of all possible interpretations of the world, meaning that it would be one of the poorest in meaning. This thought is intended for the ears and consciences of our mechanicists who nowadays like to pass as philosophers and insist that mechanics is the doctrine of the first and last laws on which all existence must be based as on a ground floor. But an essentially mechanical world would be an essentially *meaningless* world. Assuming that one estimated the value of a piece of music according to how much of it could be counted, calculated, and expressed in formulas: how absurd would such a "scientific" estimation of music be! What would one have comprehended, understood, grasped of it? Nothing, really nothing of what is "music" in it![3]

The judgment is harsh and the words caustic. But philosophers do not have a monopoly on such opinions. Scientists have plenty of reasons of their own for wanting to promote metaphysics at the expense of scientism. Erwin Schrödinger, for one, confessed similar reservations: "Most painful is the absolute silence of all our scientific investigations towards our questions concerning the meaning and scope of the whole display [of the physical world]."[4] Far from being a prelude to science, metaphysics actually constitutes its horizon. Kant understood well the imprint metaphysical ideas leave on man's scientific endeavors. Schopenhauer would later expand on the theme: "*Physics* is unable to stand on its own feet, but needs *metaphysics* on which to support itself, whatever fine airs it may assume towards the latter.... The *physical* explanation, in general and as such, still requires one that is *metaphysical*, which would furnish the key to all its assumptions."[5]

Just the same, unifying the four fundamental interactions remains a great ambition of physicists, a noble dream that must be treated with all the respect it deserves. One should not dismiss with condescending impatience the enormous progress achieved of late. Much like paleontologists are able to reconstruct an entire skeleton from a single tooth, theorists strive to imagine the phenomenal energies at the very first instants of the universe, when three of the four forces were, by all evidence, undifferentiated.

A "Grand Unification"?

Future accelerators will perhaps confirm the Higgs mechanism for the unification of the first two interactions. But the search for unification will not stop there. Particle physicists believe it is possible to devise a scenario bringing together not only the electromagnetic and weak interactions, which have already been unified, but the strong nuclear interaction as well. In anticipation, they have even already come up with a name for the theory—they call it the "grand unification." They are following much the same path that succeeded before, which consists in looking for symmetries broader than the ones that apply to the electroweak and the strong interactions separately. Despite some early promising leads, the trail has grown somewhat cold in the last decade.

The strong interaction was also adequately described in the framework of gauge theory. Physicists do not claim to have unified it with the others, but they have been able to satisfactorily account for all these interactions using a theoretical formalism similar to quantum field theory. The common description of these three interactions—the electroweak interaction, which combines weak interactions and electromagnetism, and the strong interaction—forms what is commonly referred to as the standard model. If nothing else, the model, thus far completely validated in the range of energies currently accessible, offers a convenient classification scheme of the constituents of matter.

As outlined in Chapter 4, fermions include leptons, which can propagate freely, and quarks, which are always bound to other quarks or antiquarks. All fermions are grouped in three distinct families. The first is comprised of two leptons (the electron and the electron neutrino) and two quarks (*u* and *d*, for *up* and *down*). According to this model of distinctly atomistic flavor, these fundamental units make up virtually everything in our normal environment, including ourselves. The members of the other two families occur only in special circumstances, such as high-energy collisions. The second family includes the muons (corresponding to the electron), the muon neutrino (corresponding to the electron neutrino), and the *c* and *s* quarks (for *charm* and *strange*). The third and last family contains the tau lepton and its matching neutrino (yet to be discovered), as well as the *b* and *t* quarks (for *beauty* and *top*). The discovery of the top quark—the last missing link—was announced on April 26, 1994, by physicists at Fermilab in Batavia, Illinois.

The standard model is based on the principle of gauge invariance, which provides an extremely elegant mathematical formulation to describe the forces of nature. Each interaction stems from a particular symmetry. Since this approach has proved so effective before, theorists hope to unify at least three of the four fundamental interactions by using similar principles. This ambition, quite legitimate in itself, is also motivated by the fact that the standard model is not entirely satisfactory. For starters, it suffers from many of the thorny problems of interpretation that have plagued quantum mechanics. In addition, it fails to provide answers to a number of fundamental questions. No one knows, for instance, why there are three families of particles, rather than just 1 or 999.

Is it possible to come up with something better than the standard model? The fact that the strong interaction can be described in the same framework as the electroweak interaction would suggest that the entire package ought to be unifiable, presumably by looking for a more generalized symmetry. This idea was proposed as early as 1972, under the somewhat unfortunate name of "grand unification." The hopes that were once riding on it have been somewhat tempered. It just does not seem that the three interactions really converge at high energies, contrary to earlier speculations. What is more, some versions of the theory imply that protons are unstable and should decay very slowly but perceptibly. Yet, not a single experiment has thus far produced any shred of evidence of proton decay. At the moment, these avenues are uncertain at best. The only promising lead on the horizon is the recently proposed concept of "supersymmetry."

A "Super" World?

We have seen that quantum theory, the accepted framework of particle physics, distinguishes two types of particles—fermions, particles that are very much loners, and bosons, a much more congenial and sociable bunch.

Fermions obey Pauli's exclusion principle, which states that no two fermions can be found in the same physical state. In accordance with this principle, it is impossible to find two fermions in a given location at the same time. This tenet reflects the fundamental impenetrability of matter and explains the hardness and cohesiveness of macroscopic objects. Bosons, on the contrary, conform to what might be called the principle of gregariousness. They "prefer" to be together and "do the same thing." They even enjoy the right of "superposition." They can appear or disappear in any number, provided it is even. Any physical process can only change the number of bosons in multiples of two.

In light of these differences, fermions are generally perceived as the genuine constituents of matter, its "elementary bricks," as it were. They come in a set of six quarks (fermions which are subject to the strong interaction) and six leptons (which are not). One might add an equal number of antiparticles, which are responsible for the existence of antimatter.

Bosons, on the other hand, are interaction (also called "exchange") particles. They serve as mediators of interactions.[6] Photons are responsible for electromagnetism, intermediate vector bosons for the weak interaction, and gluons for the strong interaction between quarks. We might add gravitons, should we ever succeed in describing gravitation in a similar formalism. Given the fact that each particle has a corresponding antiparticle, we end up with a total of about forty particles. Some physicists are of the opinion that this number is far too high, convinced as they are that truly fundamental particles have to be very few in number. In their view, the present classification scheme is too prolific: If it takes three families of fermions, that is obviously two too many.

All hope to restore some unity is not lost, though. To start with, even in the absence of any theoretical or experimental proof, it is not unreasonable to assume that the particles known today are actually composites, and that their eventual description (which remains to be discovered) will involve a smaller number of new and truly elementary constituents. We might even imagine that the quest of elementarity will never be consummated, and that we will keep on discovering that objects believed to be elementary at a given time actually have an underlying composite structure. In a certain sense, this idea is completely at odds with the very principles of atomism.

Another and very different point of view holds that there is an infinite variety of elementary particles, of which only a very small number can be observed at the energies reachable with current technology. This proposition is the basis of what is known as "superstring theory," which describes particles in terms of long linear objects, structured somewhat like spaghetti strings in a multidimensional space.

The differences between fermions and bosons hint at a very profound asymmetry in physics, one that appears much more fundamental than the asymmetry differentiating a neutron from a proton, or even a proton from an electron. Yet, even if "man cannot repair what God tore in two pieces," in the words of the physicist Wolfgang Pauli, it is not etched in stone that this duality will not in due time be unified as well. It would be an immense source of satisfaction to find out that the distinction between particles and interactions can disappear altogether in favor of a new and more all-encompassing concept. The future will tell whether such a concept can emerge from *supersymmetry*.

Supersymmetry

Many theorists take the view that modern physics obeys a broader fundamental symmetry named *supersymmetry*. As we have seen, unification has often come about in the past as a result of introducing new types of symmetries.

As it happens, supersymmetry possesses extremely attractive properties from this point of view. First, it establishes a correspondence between fermions and bosons, in that they would simply be "mirror images" of each other. Supersymmetry also has the virtue of generalizing the fundamental symmetries of space and those of space-time familiar in special relativity.

Any symmetry can be understood, as we have pointed out, as an invariance with respect to a certain type of transformation. In the case of supersymmetry, the relevant transformation changes a fermion into a boson, and conversely. It is not a purely geometrical operation, as it can only be expressed through a rather cumbersome mathematical formalism. Nonetheless, it has the remarkable property that two such transformations applied in succession are equivalent to an ordinary translation. This result suggests that supersymmetry might be even more fundamental than an invariance by translation. The latter might simply be a consequence of the former.

This is a potentially fundamental discovery. There are indications that supersymmetry might be able to "salvage" the grand unification theory. The reason has to do with one of the consequences of supersymmetry, namely, that the three interactions of interest would converge at an energy of about 10^{16} GeV. If so, the earlier predictions of proton decay would be nullified, and the grand unification theory might prove consistent with experimental data after all.

What makes supersymmetry even more interesting is how closely it appears to be related to space-time and its geometrical transformations. This could pave the way for an even broader unification that would incorporate gravitation into what is called *supergravity* theory (see below).

Supersymmetry represents a major conceptual breakthrough in that it eliminates the distinction between entities subject to an interaction—fermions—and the agents of that interaction—bosons. The two were already treated on a relatively equal footing by quantum field theory, and the novelty of the concept is not all that revolutionary. Many view it as just another unification, perhaps one that is necessary to move physics forward.

The problems have more to do with the extreme complexity of the mathematical treatments. As it turns out, there are several possible variations of the model, which have yet to be sorted out. A kind of outline of a supersymmetrical standard model has actually been worked out, in which each known particle would be associated with another new hypothetical particle, which would be its "supersymmetrical partner." None of these new particles has ever been observed, but there are experimentalists who are already formulating plans to look for them with the next generation of particle colliders. At the same time, astrophysicists are wondering whether these particles may hold the key—or at least a partial answer—to the puzzle of "the missing mass in the universe."[7]

The Fabric of the World

Why does the world have four dimensions (three in space and one in time), and not more? If there is no clear answer to the question, it is perhaps because the true number of dimensions is actually ten or twenty-six. In the 1930s, Kaluza and Klein had already proposed a space-time with five dimensions. Today, some physicists are pursuing similar ideas, resorting to an even larger number of dimensions, as well as a radically new concept. Fundamental objects would no longer be particles (with zero dimension), but long linear and unidimensional entities visualized as vibrating strings. That is the general principle underlying what has come to be known as string theory.

One of the motivations of those who are working on such a theory is to get rid of troublesome singularities in field theory calculations. Indeed, catastrophic divergences keep turning up the moment one deals with interactions on very small scales. Quantum theory gets around these difficulties with a method that is as artificial as it is effective; it is called renormalization.[8] But the procedure is quite arbitrary and lacks any good justification. Furthermore, it does not apply to gravitation. It appears that the concept of string might help solve the problem, since it deals from the outset with the structure of space and elementary objects on a very small scale (called Planck's scale). And, sure enough, problems of singularities show up in a totally different way in that theory.

Unfortunately, there is as yet no fully developed string theory. There is only a set of equations that make it possible—in the view of some experts—to describe what the dynamics and evolution of these exotic objects might look like. No principle, either scientific or metaphysical, is guiding physicists down this road.

There are several possible ways to put together a string theory. To begin with, in order to be workable, it must be related to supersymmetry theory and its natural outgrowth—superstrings. Generally, things take place in a space-time manifold with ten dimensions, six of which are folded back onto themselves so as to give the appearance of our traditional four-dimensional space-time. One difficulty with this approach is that there are too many ways to envision how the extra dimensions can be disposed of. Nevertheless, recent progress suggests that these difficulties might be resolved with the concept of "duality." This is yet another fundamental mathematical concept, which is not quite a symmetry but looks very much like one, since it establishes correspondences between mathematical objects. It is the ultimate irony that something called "duality" may turn out to be the price to pay to ensure the unity of physics.

A Theory of Everything?

Interest in the grand unification remains relatively limited in the physics community. Whether or not a unified vision of subnuclear processes can be worked out is of no immediate concern to the rest of physics. Of far greater importance would be the incorporation of gravitation in such a scheme or, to be more precise, the unification of the four interactions in a common formalism, whatever that may turn out to be.

It seems crucial to try to unite at least gravitation and electromagnetism. That remains the most important problem faced by physics today. Einstein is the one who originally issued this ambitious challenge; no one has been able to meet it thus far. The task is formidable. Indeed, unifying the first three interactions was no breeze, despite the fact that they fit in the common framework of quantum theory. But gravitation has up to now refused to be tamed by quantization, for reasons that seem increasingly insurmountable. This apparent incompatibility between gravitation and quantum mechanics has no practical importance to speak of. In general, gravitation applies to large spatial scales typical of astronomy, where quantum effects are inconsequential. To each his own domain, and everyone is happy! But the problems are enormous on a conceptual level. Moreover, they come up in concrete ways as soon as one is interested in the very early moments of the universe. Densities and energies were so large then that relativistic and quantum effects had to be present simultaneously. The holy grail is to at least reconcile the two, if not unify them.

How to reach that goal once seemed quite straightforward. Since the first three interactions can be described in terms of quantum physics, it seemed logical to "quantize" gravitation as well before attempting to incorporate them all in a common scheme. With a little bit of luck, unification might even pop out as a bonus. It sounded like a good plan, except for the fact that every attempt to execute it has so far hit a brick wall. Quantizing gravitation probably means that both geometry itself and time must be quantized as a prerequisite. As it happens, "ordinary" quantization operates in predetermined space and time. It does not deal specifically with space and time, which are merely a passive backdrop for events to unfold against. This creates enormous difficulties for their quantization. The only way to circumvent the problem may be to adopt a cosmological approach from the outset. For this reason, the primordial universe (including black holes, if they do exist) constitutes an ideal testing ground for any quantum theory of gravitation.

Aside from certain notions related to quantum cosmology, which we will discuss shortly, the most promising lead currently for incorporating gravitation in a unified scheme exploits the concept of supersymmetry.

Supergravity

Supersymmetry is a type of symmetry that can be qualified as global. This term has a specific meaning in the language of quantum field theory. As it happens, the quantum theory of electromagnetism, widely considered one of the great triumphs of twentieth-century physics, can be interpreted as resulting from an initially global symmetry transformed into one that is local. At least, that is the rationale behind gauge theory. Can the same feat be duplicated with supersymmetry? Is it possible to construct a theory in which supersymmetry has the attributes of a gauge theory? That would have exciting repercussions because gravitation itself can be considered a theory of just such a type (related as it is to the Lorentz transformation).

Some theorists are diligently trying this approach to construct a supergravity theory that would incorporate and extend the basic ideas of supersymmetry. As the name implies, supergravity would encompass gravitation by unifying it with the other interactions. Some evidence suggests that singularities might disappear entirely, obviating the need for any renormalization.

For all its appeal, the idea remains unfortunately a very tough challenge, conceptually as well as practically. Calculations are so complex that it is virtually impossible to hazard a prognosis about the chances for success.

Quantum Cosmology

It is probably safe to predict that the theory of everything is not around the corner. Even the less ambitious grand unification is not visible on the horizon just yet. The proposed paths prove horrendously forbidding. Thus far, they have failed to provide any new answer to fundamental problems concerning the nature of space and time. In fact, all indications are that if one insists on quantizing gravitation, the only viable solution is probably to devise a quantum cosmology.

"Ordinary" quantum physics (which excludes gravitation) leads quite naturally to cosmology. Indeed, we have discussed how the celebrated principle of "quantum nonseparability" precludes two or more objects that have interacted in the past from being treated as separate entities. Obviously, most objects in the universe have had some mutual interaction, if only through gravitation or radiation exchange. As a result, even though quantum physics has proved eminently suitable to describe isolated objects with great accuracy, strictly speaking, it can be correct only if it considers the entire universe as a single physical system. From a practical standpoint, though, it is impossible to comply with that provision because quantum theory is not amenable to treating space as a dynamical variable, as required by cosmology. That makes it all the more imperative to devise a quantum version of cosmology, if the consequences of the principle of nonseparability are to be taken seriously.

A few physicists are devoting their attention to the wave function of the entire universe. The difficulty they run into is that it must encompass all the characteristics of the universe, including space and time.[9] If so, it is evidently no longer possible to consider that things evolve *in* space and *in* time. These fundamental problems are perhaps the most serious obstacles to constructing a quantum cosmology. But any attempt to improve physics will sooner or later be confronted with these very issues.

Toward the end of the 1970s, the American physicists John Wheeler and Bryce de Witt proposed a model that might put us on the right track toward a genuine quantum cosmology. Their theory was given the name *Quantum Geometrodynamics*. Its goal is to describe the universe from both a quantum and a relativistic perspective. It effectively catalogues all possible models of the universe in terms of their instantaneous states at a given moment of their evolution. For instance, the big bang model enters only through its present state.[10] With this approach, temporal problems are alleviated and the theory is considerably simplified.

Needless to say, it is a full-fledged quantum theory. At any given moment, the universe does not find itself in any particular state defined in the classical sense, not anymore than a quantum particle occupies a specific position in space. A quantum state of the universe is described by a wave function that is a "quantum superposition" of classical states, each of which corresponds to what we would normally call a universe in a relativistic sense.[11] The evolution of the wave function is described by a single equation known as the Wheeler-de Witt equation, which amounts to a generalization of Schrödinger's equation.

After injecting a few additional hypotheses into the model, it becomes fairly painless to carry out some calculations. Interpreting the results is quite another matter. Certain configurations of the universe (for instance, those in which space is finite, or its topology complex) turn out to be more probable than others. Unfortunately, theorists are having a difficult time coming up with satisfactory interpretations of the probabilities they have learned to calculate.

Many questions remain unanswered: What is the proper way to interpret a superposition of states? Why does the universe give the appearance of being in a specific (classical) state if it is actually a quantum superposition? What does the notion of probability mean in such a context? Most of these questions are in fact essentially the same as those that have dogged ordinary quantum physics as applied to particles. They are simply rephrased in a cosmological language. The general consensus is that the geometric state of the universe is not fixed in a quantum cosmological world anymore than the position of a particle is in quantum mechanics. The universe might be in several distinct states in the classical sense "at the same time."

Even though a quantum version of cosmology is nowhere near perfected, some have already attempted to apply it. The physicists James Hartle and Stephen Hawking have pointed out that the relevant equations, like those in many other branches of physics, can only be solved by specifying the "boundary conditions" of the universe.[12] That introduces an element of arbitrariness in cosmology that not everyone is comfortable with. Hartle and Hawking in fact tried to develop a model without any boundary condition imposed on either space or time. In such a picture, the problem of the initial singularity disappears, and the universe has no beginning or end.[13]

The Russian physicist Andrei Linde proposed an entirely different idea. He assumed that the initial conditions were "chaotic," which leads to a solution in the form of an enormous universe regenerating itself endlessly. It is oftentimes likened to a "foam of mini-universes," of which each bubble would have its own characteristics, including its physical constants, number of dimensions, and dynamical laws. Our immediate surroundings, which we perceive as our own observable universe, would be only an infinitesimally small part of just one of these innumerable bubbles. The precise meaning of the unity of the world is not altogether clear in such a model. Some might even object that a fragmented unity is no unity at all.

Conclusion

The human mind feels a special connection with unity. The discovery process is imbued with monism. Man's yearning for intelligibility simply cannot do without the notion of the One. Yet, that does not guarantee that a methodological search for unity in knowledge can proceed unbiased by an ontological certitude that the world must be one. It is not enough to acknowledge such a compelling predisposition in human nature to validate everything it produces. Unity could well prove to be an illusion after all. It may turn out to be nothing but a mirage induced by a metaphysical obsession.

Long before the advent of physics, many philosophers of nature had tried various approaches to forge a global and unified view of the world. At the same time, the proliferation of such attempts is in itself persuasive evidence that none really succeeded. Our understanding of things remains very disjointed. Because they are devised and conducted by humans, experiments are inherently deficient. Unity, a constantly changing intellectual construct, is elusive. It takes immense patience and determination to flush it out of hiding.

Science is reductionistic by essence. But it is necessary to make a distinction between methodological reductionism and ontological reductionism. The first is desirable because any science can progress only by streamlining its explicative principles and by looking for the unity hidden in its models. The second is much less rewarding because it is tainted by a dogmatic faith that everything emanates from some supreme existent that science has the power to grasp.

Physics came into being through a combination of rudimentary syntheses and unifications. It began with a mechanistic model that tied together matter, the heavens, astronomy, dynamics, and geometry. We have seen how various doctrines overcame sometimes profound contradictions to give us the Newtonian synthesis of harmony and atomism; the integration of motions on earth and in the heavens; and the unification of matter, of space, and ultimately of the entire world. Perhaps the most crucial development was the concept of the universe itself, which ensured the universality of all physical laws.

The search for unifications continues to be the guiding principle of physics to this day. Harmony, atomism, teleology, geometrization, sundry analogies—everything is tried, discarded, and occasionally rehabilitated. The growth of knowledge results from successful consolidations and unifications of separate disciplines, notwithstanding the many failures littering its path.

Yet, modern physics is anything but unified; it appears fragmented and compartmentalized. The prospects for an all-encompassing unity are questionable at best. Indeed, it could be argued that we have never been further from a globally unified worldview, and that is a frustrating paradox.

We are faced with an apparent contradiction. On the one hand, physics is manifestly fueled by a unifying motive. Unity often results from the discovery of new concepts or hypotheses that, almost overnight, overthrow multiplicity and install homogeneity in its place. On the other hand, the process never seems to reach its culmination. Either it cannot be pushed to its ultimate conclusion, or it succeeds only too briefly until new results bring fresh elements into the picture and force a complete reassessment of where we stand. The breadth of each discipline keeps increasing and the landscape of knowledge shifts constantly. Two branches are no sooner unified than a third begins to grow. When, in turn, it is merged with the first two, still more appear. The cycle is endless, and no magic formula can ever fundamentally change that. Will we be able to reach higher levels of understanding leading to new unifications? Probably, and it is safe to predict that it will be at the cost of further setbacks to the cause of unity. At best, partial unifications strengthen our resolve to keep searching for further breakthroughs.

If physics works at all, it is perhaps because such unifications are amenable to mathematical treatments and formulations, which become useful tools to describe, if not the world itself, at least all the matter it contains. This process—almost dialectical in nature—is absolutely vital. Should it ever stop, science would simply wither.

Under these circumstances, should we simply give up looking for the underlying unity of diverse phenomena? The answer is a resounding no—that would be tantamount to giving up on physics itself. Yet, the irony is that unification does not necessarily lead to unity. It weaves a fabric that is never quite finished. Rarely does it offer more than a fleeting glimpse of unity. Interim stages are a quintessential part of science and its main stimulus; they constantly provide new incentives to press on for a better understanding of things, more novel theories, and more creative insights.

Modern physics has scored remarkable triumphs. That exposes it to the risks that inevitably come with success. In its eagerness to publicize its next anticipated achievement, it can easily degenerate into metaphysical contemplations—some are beneficial because they inspire bold hypotheses, but others are far more insidious and prone to lure it into arrogant overconfidence that it is within reach of its destiny. Only if it renounces pretensions to be a perfect reflection of reality will physics continue to thrive. Its greatest challenge is to resist the temptation to project the false self-image of a religion able to reveal the ultimate truth.

Notes

Introduction

1. See Gottfried W. Leibniz, *New Essays on Human Understanding*, trans. Peter Remnant and Jonathan Benett (New York: Cambridge University Press, 1981).
2. Pierre Duhem, *The Aim and Structure of Physical Theory*, trans. Philip P. Wiener (New York: Atheneum, 1962), p. 103.
3. Ibid., pp. 103-104.
4. Antoine A. Cournot, *An Essay on the Foundations of Our Knowledge*, trans. Merritt H. Moore (New York: Liberal Arts Press, 1956).
5. Plato, *Timaeus*, 55d, trans. Desmond Lee (New York: Penguin Books, 1965).
6. Ibid., 53e.
7. Ibid., 28a.
8. Ibid., 69a.
9. In the context of discussing the process of vision, the formation of images in mirrors, and mechanical causes in general, Plato observed: "All these are among the adjuvant causes, which God uses in shaping things the best way possible. But they are thought of by most people not as accessory but as true causes, achieving their effects by heat and cold, solidification and liquefaction, and the like." Ibid., 45c.
10. Ibid., 48a. Necessity conceived of in this way has no final design. It is what Plato referred to as the "errant" cause.
11. Antonia Soulez, ed., *Manifeste du Cercle de Vienne et autres écrits* (Manifesto of the Vienna Circle and other writings) (Paris: Presses Universitaires de France, 1985).
12. This caused Pierre Duhem much aggravation (see Duhem, *The Aim and Structure of Physical Theory*). Duhem recognized Maxwell's theory as original,

and he disapproved of any attempt to describe it in terms of concrete models, a trend he blamed Maxwell himself for.

13. Stephen W. Hawking, *A Brief History of Time* (New York: Bantam Books, 1988), pp. 174-175.
14. See, for instance, Paul K. Feyerabend, *Farewell to Reason* (London, New York: Verso, 1987).
15. Some mathematicians, such as Georg Cantor, set themselves the goal of unifying mathematics in order to reach "the ultimate unity that makes it possible to consider from the same point of view the continuous and the discontinuous, and to measure them with a common yardstick" (quoted by Jean-Luc Verley in *Georg Cantor*, Encyclopedia Universalis, vol. 4 [Paris: Corpus, 1988]). The process can lead to stunning surprises, even for those making the discoveries. Having just established a correspondence between a one-dimensional continuum and one with several dimensions, or, put another way, between points of a line and those of a plane, volume, and so on, Georg Cantor wrote to his friend Julius Dedekind: "I can see it, but I don't believe it."

Chapter 1

1. Wilhelm Ostwald, "La Déroute de l'atomism contemporain" (The disarray of contemporary atomism), *Revue générale des sciences pures et appliquées* 21 (1895): 953-958.
2. See Dominique Lecourt's commentaries in *À quoi sert donc la philosophie? Des Sciences de la nature aux sciences politiques* (What is good for philosophy? From the sciences of nature to political science) (Paris: Presses Universitaires de France, 1993), pp. 92-93.
3. Aristotle defines the object of *phusis* thus: "The nature of a thing, then, is a certain principle and cause of change and stability in the thing, and it is *directly* present in it—which is to say that it is present in its own right and not coincidentally." See Aristotle, *Physics*, Book II, $192^{b}20$, trans. Robin Waterfield (New York: Oxford University Press, 1996).
4. Friedrich Nietzsche, *Philosophy in the Tragic Age of the Greeks* (Chicago: Henry Regney Co., 1982).
5. Ernst Cassirer, *The Philosophy of Symbolic Forms*, vol. 2, *Mythical Thought*, trans. Ralph Manheim (New Haven, Conn.: Yale University Press, 1957).
6. As Leibniz was to say later: "When a rule is too complex, what conforms to it appears to lack any rule." See Gottfried W. Leibniz, *Discourse on Metaphysics*, trans. Daniel Garber and Roger Ariew (Indianapolis, Ind.: Hackett Publishing Co., 1991).
7. This does not change the fact that some explanations actually amount to popular myths.
8. See, for instance, Gilles Deleuze, *Nietzsche et la philosophie* (Nietzsche and philosophy) (Paris: Presses Universitaires de France, 1962), or Clément Rosset, *Le Choix des mots* (The choice of words) (Paris: Éditions de Minuit, 1995).
9. Jean de la Fontaine, *Fables* (Paris: Imprimerie Nationale, 1985). The fable in

question, entitled "Les deux rats, le Renard, et l'Oeuf" (The two rats, the fox, and the egg) is in Book 9.

10. Jean Voilquin, ed. and trans., *Les Penseurs grecs avant Socrate. De Thalès de Milet à Prodicos* (Greek thinkers before Socrates: From Thales of Miletus to Prodicos) (Paris: Garnier-Flammarion, 1964), section 61.
11. In the late 1920s, the British physicist Paul Dirac was trying to integrate Einstein's laws of special relativity, which apply to objects whose speed approaches that of light, into the then-new quantum mechanics, which he helped found. In 1927, he came up with an equation that has come to be known as the relativistic Dirac wave equation. It provided a perfect description of electrons and made it possible to determine with improved accuracy the energy levels of a hydrogen atom. But it also presented a puzzling problem in that half its solutions seemed to correspond to kinetic energies that were negative. That seemed to make no sense, since the kinetic energy of a body in motion is always a positive quantity. What did these solutions mean? After trying several hypotheses, Dirac proposed that they corresponded to particles whose energy was indeed positive. Their mass would be identical to that of electrons, but they would carry an opposite charge. He called them antielectrons (they were later renamed positrons). Dirac turned out to be correct. As early as 1932, the American physicist Carl Anderson detected a few positive electron twins in the shower of particles produced by cosmic radiation plowing through the atmosphere. It is a general property of relativistic quantum mechanics that each particle has an antiparticle counterpart of similar mass and opposite charge, the two being in effect mirror images of each other.
12. Cited in Roland Jaccard, *Le Cimetière de la morale* (The graveyard of morality) (Paris: Presses Universitaires de France, 1995), p. 14.
13. Werner Heisenberg, *Physics and Philosophy: The Revolution in Modern Science* (New York: Harper & Row, 1962), p. 62.
14. Plato, *Sophist*, 259d-e, trans. Seth Bernadeto (Chicago: University of Chicago Press, 1984).
15. A less well-publicized account of his death has it that he simply hanged himself.
16. Both philosophers, incidentally, were perceived by Athenians as dangerous rebels. For having asserted that the moon was simply a rock (rather than a goddess), Anaxagoras found himself threatened with charges of impiety. Socrates was accused of corrupting the youths and offending the gods of the city. He had a less fortunate fate, as history records with the infamous episode of the poisonous hemlock drink.
17. Anaximander, in *Les Présocratiques*, (The Pre-Socratics), ed. J.-P. Dumont (Paris: Gallimard, "Bibliothèque de la Pléiade," 1988), p. 39.
18. Parmenides, in ibid., p. 243.
19. Ibid., p. 258.
20. Henri Poincaré, "Les Rapports de la matière et de l'atome" (The relationships between matter and the atom), in *Dernières Pensées* (Last thoughts) (Paris: Flammarion, 1963 [1913]).

21. Paul Dirac proposed just such a theory. Unfortunately, various cosmological observations have established that these "constants" are indeed constant over extremely long periods of time.
22. Gaston Bachelard, *Les Intuitions atomistiques. Essai de classification* (Atomistic intuitions: An attempt at classification) (Paris: Vrin, 1975 [1935]), p. 102.
23. The concept of atom did not meet with immediate acceptance. Several years after Jean Perrin had experimentally demonstrated its existence, the renowned French thinker known as Alain (whose real name was Emile Chartier) wrote: "Atomism is for morons. In the eyes of great geniuses, it is merely a convention." See Alain, *Histoire de mes pensées* (A history of my thoughts) (Paris: Gallimard, 1936), p. 47.
24. The history of the atom is not without irony. The model originally proposed in 1913 by Niels Bohr in a certain sense vindicated the energeticists, who put energy before matter and were in fact opposed to atoms. In Bohr's model, electrons were characterized not by their granularity, position in space, or speed, but simply by the energy level corresponding to the stationary states they happen to be in.
25. See Lucretius, *The Nature of Things*, Book II, 293, trans. Frank O. Copley (New York: W.W. Norton & Co., 1977).
26. Gottfried W. Leibniz, *Oeuvres* (Collected works) (Paris: Éd. L. Prenant, 1972), p. 473.
27. The reader is referred to the excellent book by Bernard Pullman, *The Atom in the History of Human Thought* (New York: Oxford University Press, 1998).
28. Plato, *Phaedo*, 97d-98c, in *Five Dialogues*, trans. G. M. A. Grube (Indianapolis, Ind.: Hackett Publishing Co., 1981).
29. Ibid., 99b.
30. Blaise Pascal, *Pensées*, trans. A. J. Krailsheimer (New York: Penguin Books, 1966), thought 199.
31. On this point, the reader is urged to consult the book by Gerald Holton, *The Scientific Imagination: Case Studies* (New York: Cambridge University Press, 1978). Einstein's conception of the essence of theory is analyzed in some detail in Chapter 3.
32. In Einstein's opinion, while both special relativity and general relativity were theories of principle, quantum physics was a "constructive" theory. He felt that, in spite of its many remarkable successes, it would someday turn out to be a limiting case of a more general theory, as yet unknown, "just as electrostatics is deducible from the Maxwell equations of the electromagnetic field or as thermodynamics is deducible from statistical mechanics" (cited in Abraham Pais, *Subtle Is the Lord: The Science and the Life of Albert Einstein* [New York: Oxford University Press, 1982], p. 461).
33. Albert Einstein, "On the Method of Theoretical Physics," *Philosophy of Science* 1 (1934): 163-169.
34. Bernard Ribémont, "Le Moyen Age et la symbolique des nombres" (The Middle Ages and the symbolism of numbers) *La Recherche*, 278 (1995): 736.
35. The Wisdom of Solomon, 11:20.
36. Kepler nested between the orbits of the six planets known at the time the

five regular polyhedra that are the centerpiece of Euclid's *The Elements*.

37. Having clarified the rules of the game, it should come as no surprise that the notes associated with earth are *mi* and *fa*, for *mi*sery and *fa*mine.
38. Arthur Koestler, *The Sleepwalkers: A History of Man's Changing Vision of the Universe* (New York: Grosset & Dunlap, 1963), p. 340.
39. Pierre Cartier, "Képler et la musique du monde" (Kepler and the music of the world) *La Recherche* 278 (1995):750.
40. For the benefit of the inquisitive reader, the result is $R = 2\pi^2 m e^4 Z^2 / h^3$, where m is the mass of the electron, e its electrical charge, Z the atomic number, and h is Planck's constant.
41. The tetrahedron stood for fire, the icosahedron for air, the cube for water, and the octahedron for earth. In addition, the dodecahedron represented ether, which was the fifth element, or "quintessence," and filled the heavens.
42. See, for instance, Loup Verlet, *La Malle de Newton* (Newton's baggage) (Paris: NRF Gallimard, 1993), or Richard S. Westfall, *Never at Rest: A Biography of Isaac Newton* (New York: Cambridge University Press, 1980).
43. Westfall, *Never at Rest: A Biography of Isaac Newton.*
44. Quoted in Michel Paty, *Einstein philosophe: La Physique comme pratique philosophique* (Einstein the philosopher: Physics as philosophical practice) (Paris: Presses Universitaires de France, 1993).
45. Albert Einstein, *The World as I See It*, trans. Alan Harris (New York: Philosophical Library, 1949).
46. This statement assumes that the phrase "true nature of reality" has a meaning. Not everyone is prepared to accept that.

Chapter 2

1. Saint Thomas Aquinas, *Commentary on the Metaphysics of Aristotle*, Book V, trans. John P. Rowan (Chicago: Henry Regnery Co., 1961).
2. Nicholas of Cusa, *Of Learned Ignorance*, trans. Jasper Hopkins (Minneapolis, Minn.: A. J. Banning Press, 1988).
3. Joseph Fourier, *The Analytical Theory of Heat*, trans. Alexander Freeman (New York: Dover, 1955). See Preliminary Discourse, p. 1.
4. Ibid., p. 2.
5. Ibid.
6. Ibid.
7. Pierre Thuillier, *La Grande Implosion: Rapport sur l'effondrement de l'Occident* (The great implosion: A report on the collapse of the West) (Paris: Fayard, 1995), p. 63.
8. René Descartes, "Letter to Mersenne," 11 March 1640.
9. Auguste Comte would later paraphrase the same idea: "There shall be no observation without an implicit theory." Gaston Bachelard insisted that there can be no experiment without the prior formulation of a problem.
10. Antoine A. Cournot, *Considérations sur la marche des idées et des événements dans les temps modernes* (Considerations on the march of ideas and events in mod-

ern times) (Paris: J. Vrin, 1975 [1872]), p. 187.

11. Alexandre Koyré, "Attitude esthétique et pensée scientifique" (Esthetical view and scientific thought) in *Études d'histoire de la pensée scientifique* (Studies in the history of scientific thought) (Paris: Gallimard, 1973), p. 129.
12. Jacques Maritain, *The Dream of Descartes*, trans. Mabelle L. Andison (New York: Philosophical Library, 1944), pp. 47-48.
13. Galileo Galilei, *Discourse on Bodies in Water*, trans. Thomas Salusbury (Urbana, Ill.: University of Illinois Press, 1960).
14. It has been debated whether the human intellect truly discovers universal and immutable laws or merely discerns the behavior and "habitus" of nature. The question goes back to antiquity, but only in the eighteenth century did it begin to be asked in the specific context of scientific knowledge, rather than knowledge in general. Bertrand Saint-Sernin makes a connection between such a question and reflections in eighteenth-century Germany on what constitutes genius: "If nature operates at all as an artist would, there is nothing to assure us that what it produces obeys rules that must be immutable in time. Why would nature not change its habits? Why would it not have its eras, fashions, and styles? Why would it not embrace new ideas? If so, it would be reason's responsibility to be alert to nature, to listen to its voice, to guess its intentions and detect its inflections. . . . If creative nature (*natura naturans*) has artistic powers, its laws are 'contingent,' then reason loses any competence of its own, and it must, in the very core of its being, accept and celebrate the arbitrary creator of nature; it must consent to inhabit a world that, as in Heraclitus's days, is 'new every day.' " See Bertrand Saint-Sernin, *La Raison au XXe siècle* (Reason in the twentieth century) (Paris: Éditions du Seuil, 1995), p. 195.
15. Richard S. Westfall, *Never at Rest: A Biography of Isaac Newton* (New York: Cambridge University Press, 1980), p. 7.
16. Yet, some of Galileo's own pronouncements seem to contradict these innovative concepts. See, for example, Koyré, in *Études d'histoire de la pensée scientifique*.
17. Galileo Galilei, *The Sidereal Messenger*, trans. Albert van Helden (Chicago: University of Chicago Press, 1989), p. 40.
18. Galileo Galilei, *Dialogue Concerning the Two Chief World Systems: Ptolemaic and Copernican*, trans. Stillman Drake (Berkeley: University of California Press, 1967), p. 59.
19. We know today that these stars are not really new. They are in fact so-called "novae" or "supernovae," which had existed all along but were too dim to be seen. Their brightness suddenly increased as a result of a gigantic explosion.
20. It also states its isotropy, meaning that all directions are equivalent.
21. Centuries earlier, Heraclitus had made the following pertinent comment: "For those who are wide awake, there is a single universe which is the same for everyone; but every sleeper drifts off into his own particular world."
22. Jacques Desmarets and Dominique Lambert, *Le Principe anthropique* (The anthropic principle) (Paris: Armand Colin, 1994).

23. F. de Gandt, "Le Statut particulier de la cosmologie dans la science" (The particular status of cosmology in science), in *Philosophie de la Nature* (Philosophy of nature); Ouvrage Collectif (Paris: Institut Catholique de Paris, 1992).
24. Roughly similar calculations had been independently presented a few years earlier by the Soviet mathematician Alexander Friedmann. But the work was primarily on a mathematical rather than physical level, and it remained largely unnoticed.
25. This required accepting an ad hoc "spontaneous creation of matter." Although not especially satisfying, the hypothesis violated neither experiment nor observations.
26. See, for instance, Marc Lachièze-Rey, "Faits et théories: I'Univers est en expansion" (Facts and theories: The universe is expanding), in *Science et Vie*, 189 (December 1994), in a special issue entitled "Le Big Bang en questions" (The big bang under scrutiny).
27. The properties of the cosmic background radiation have recently been measured with considerably greater accuracy. They entirely confirm the models. The same is true of the abundance of the various chemical elements in the universe. For more details on this topic, see, for instance, Marc Lachièze-Rey, *Initiation à la cosmologie* (Initiation to cosmology) (Paris: Masson, 1996).
28. Maritain, *The Dream of Descartes*.
29. René Descartes, *Rules for the Direction of the Mind*, trans. Elizabeth S. Haldane and G. R. T. Ross (Chicago: Encyclopaedia Britannica, 1952), rule 1.
30. Maritain, *The Dream of Descartes*.
31. René Descartes, *Discourse on the Method*, trans. F. E. Sutcliffe (New York: Penguin Books, 1968), pp. 35-36.
32. Maritain, *The Dream of Descartes*, pp. 24-25.
33. Auguste Comte, *Cours de philosophie positive* (Lectures in positive philosophy) (Paris: Bachelier, 1830-1842), vol. 3. Translator's note: This is a massive treatise covering a total of 6 volumes. A "condensed and freely translated" English version (still over 800 pages long!) appeared as early as 1855 and has been reissued: See Auguste Comte, *The Positive Philosophy*, trans. Harriet Martineau (New York: AMS Press, 1974).
34. Westfall, *Never at Rest: A Biography of Isaac Newton*, p. 8.
35. Ibid.
36. Ibid., p. 15.
37. Ibid., p. 14.
38. Ibid., p. 18.
39. Ibid., p. 19.
40. Ibid.
41. Alexandre Koyré, *Newtonian Studies* (Cambridge, Mass.: Harvard University Press, 1965), p. 23.

Chapter 3

1. The first measurements of the speed of light were conducted in 1676 by the Danish astronomer, Ole Christensen Roemer, by timing the occultations of

the Galilean satellites of Jupiter viewed from earth. He correctly concluded that irregularities in the timing were due to the varying distance between Jupiter and earth.

2. Louis de Broglie, *Matter and Light; the New Physics*, trans. W. H. Johnston (New York: Dover, 1955).
3. It took a veritable stroke of genius for Maxwell to introduce—without the benefit of any experimental data to guide him—the notion of "displacement current," which was at the basis of his new theory. Perhaps he too was inspired by a search for harmony. It was to make his equations more symmetrical that Maxwell came up with the idea of adding an extra term in his equations. As it turned out, it proved to be the key to this stunning unification of electromagnetism and light.
4. In all fairness, some experiments conducted by Faraday in the middle of the nineteenth century had provided some early hints of a link between light and magnetism. As early as 1845, Faraday had demonstrated that the plane of polarization of light rotates in the presence of a magnetic field: "Therefore, magnetic forces and light are mutually related," he rightfully concluded. On the other hand, Faraday was less fortunate in his attempts to unify gravity and electricity. He desperately tried to detect electrical currents in solenoids dropped from the top of a terrace, but to no avail.
5. Niels Bohr, *Atomic Physics and Human Knowledge* (New York: Wiley, 1958), p. 70.
6. This remains true even though it has since been shown that relativity can be interpreted without recourse to geometry.
7. René Descartes, "Lettre à Plempius" (Letter to Plempius).
8. Hermann von Helmholtz, *Schriften zur Erkenntnistheorie* (Essays on the theory of knowledge) (Berlin: J. Springer, 1921).
9. René Descartes, *Treatise of Man*, trans. Thomas Steele Hall (Cambridge, Mass.: Harvard University Press, 1972), p. 113.
10. An analysis—denunciation may be a more accurate word—of all these excesses can be found in the comprehensive work by Pierre Thuillier, *La Grande Implosion: Rapport sur l'effondrement de l'Occident* (The great implosion: A report on the collapse of the West) (Paris: Fayard, 1995).

Chapter 4

1. Isabelle Stengers, "Mécanique quantique: La Fin d'un rêve" (Quantum mechanics: The end of a dream), *Cosmopolitiques* vol. 4, (Paris: Éditions la Découverte, 1997), p. 72.
2. A. Einstein, B. Podolsky, and N. Rosen, "Can Quantum Mechanical Description of Physical Reality Be Considered Complete?" Physical Review 47 (1935): 777-780.
3. Bohm first proposed his theory in 1952. It has the ability to reproduce the results of quantum physics. Its premise is to view particles as genuine corpuscules guided in their motion by a wave. This wave acts as an informational field, which would determine the particle's trajectory, not unlike a

radio message ordering a ship to change course.

4. In classical physics, when an object with finite mass and volume rotates about itself, it has what is called an angular momentum proportional to its mass and its angular rotation speed. The spin is the quantum equivalent of the angular momentum. But this analogy has limitations. Particles described by quantum physics are quite different from ordinary ones, and they should not be thought of as tiny spinning tops. That would lead to contradictions. While angular momentum in classical physics can change continuously, that is not true of the quantum spin, which can only take on discrete values. For instance, the spin of an electron can be equal to only +1/2 or −1/2 (in units of $h/2\pi$, where h is Planck's constant) when projected on any arbitrary axis. These two values correspond to rotations in opposite directions.
5. Bosons contravene one of the pillars of Leibniz's philosophy, known as the "principle of identity of indistinguishable objects," which holds that two real beings can never be identical. See Gottfried W. Leibniz, *New Essays on Human Understanding,* trans. Peter Remnant and Jonathan Benett (New York: Cambridge University Press, 1981). There are cases where N bosons can be in the same quantum state without being a single entity. The number N of particles involved is a "measurable" quantity, even though the particles themselves are indistinguishable.
6. The atom of modern physics bears almost no resemblance to the original Greek concept. In particular, contrary to the ancient model, it is not undissociable.
7. According to the theory, protons and neutrons are composed of quarks. A proton is made of two u (for *up*) quarks and one d (for *down*) quark, while two d quarks and one u quark make up a neutron. Quarks have electrical charges that happen to be fractional in terms of the charge of a proton. That charge is +2/3 for a u quark, and −1/3 for its d counterpart. This readily accounts for a total charge of +1 for a proton and 0 for a neutron.
8. Protons and neutrons are perhaps the most well-known hadrons, since they are the constituent elements of atomic nuclei, in other words, of all ordinary matter. Other hadrons include the pion, also called pi-meson, which protons and neutrons exchange continually inside nuclei. In excess of 350 hadrons have been detected either in cosmic radiation or in high-energy experiments conducted in large particle accelerators. Some are electrically charged; others are not. All are unstable, with the exception of the proton. Their lifetime is exceedingly short (less than 10^{-20} second in some cases). Hadrons fall in two distinct categories, depending on the value of their spin. Hadrons with half-integer spins (1/2, 3/2, etc.), which include protons and neutrons, are called baryons. Those with integral spins (0, 1, 2, etc.), such as pions, are called mesons.
9. See, for instance, Abdus Salam, *Unification of the Fundamental Forces* (New York: Cambridge University Press, 1990).
10. Henri Poincaré had insisted that mathematical physics was entitled to shake off the yoke of excessive logic: "One should not strive to avoid contradiction at all cost but, rather, learn to take advantage of it. Two contradictory theo-

ries can both be useful tools of research, as long as they are not mixed indiscriminately or asked to reveal the essence of things." This quote is from the introduction to Henri Poincaré, *Electricité et optique* (Electricity and optics), Part I, *Les Théories de Maxwell et la théorie électromagnétique de la lumière* (Maxwell's theories and the electromagnetic theory of light) (Paris: G. Carre, 1890-1891).

11. An electron-Volt (abbreviated as eV) is a unit of energy commonly used by particle physicists. It corresponds to the energy acquired by an electron accelerated by a potential difference of 1 volt. If the electron is initially at rest, this acceleration imparts to it a speed of 600 km/s. One Mega-electron-Volt (MeV) is equal to 1 million (10^6) eV, 1 Giga-electron-Volt (GeV) to 1 billion (10^9) eV, and 1 Tera-electron-Volt (TeV) to a 1,000 billion (10^{12}) eV.
12. One fermi is equal to 10^{-15} meter, or one-millionth of a billionth of a meter. It is roughly the size of an atomic nucleus.
13. The word "color" has a very specific meaning in particle physics. It refers to one of the attributes of elementary particles, such as quarks, and has nothing to do with, say, the color of a paint. It is simply a convenient way to describe one of its characteristics. "Color" is to the strong nuclear interaction what electrical charge is to electromagnetic interactions. However, while there are only two types of electrical charge—positive and negative—quarks have a choice of three colors, arbitrarily designated red, blue, and green (or a more patriotic red, white, and blue in the United States). That is how the theory of strong interactions acquired the name *quantum chromodynamics*. Without explaining the details, the theory postulates that individual quarks can never be observed in isolation. They only occur in groups. Either a quark combines with an antiquark to form a meson, or three quarks associate to create a baryon, such as a proton or neutron. These combinations are always "colorless." A meson resulting from, say, a red quark and an antiquark of the same anticolor is colorless by construction. Likewise, a set of three quarks, one of each color, jostling inside a proton does not have any color (perhaps the picture of a spinning wheel with multicolor bands will help).
14. In the early 1930s, beta radioactivity was something of a mystery. The energy of ejected electrons can take on any value over a wide range, which is inconsistent with the assumption that the nucleus decays in two components only. Pauli and (a little later) Fermi independently postulated the simultaneous emission of a third, massless particle, which they called *neutrino*. This elusive particle is subject to the weak interaction only. Conceived of almost as an artifact, it was not detected until much later, in 1955 to be exact.
15. CERN is a French acronym for Centre Européen de Recherches Nucléaires (European Center for Nuclear Research). Founded in 1951 and located in Geneva, Switzerland, it is a leading center for research on particle physics.
16. Gerald Holton, *The Scientific Imagination: Case Studies* (New York: Cambridge University Press, 1978).
17. Gilles Châtelet, *Aspects philosophiques et physiques de la théorie des jauges* (Philosophical and physical aspects of gauge theory) (Paris: Université Paris-Nord–IREM, 1984).

18. The standard model describes interactions by resorting to the principle of gauge invariance, as we have pointed out. One of the consequences of this principle is that all interaction particles must have zero mass.
19. It leads to the introduction of yet another field, one that is electrically neutral and subject to the weak interaction only. This so-called "Higgs field" supposedly would couple to the intermediate W^+, W^-, and Z^0 bosons, providing each of them with its mass. But it would be exempt from coupling to photons, thereby preserving their zero mass.
20. The construction of a linear electron-positron collider is also being considered, although for the more distant future. Its energy would be of the order of only 500 GeV (or 0.5 TeV), significantly less than in the LHC proton collider. However, protons are not necessarily the best choice as projectiles, since the energy released during their mutual collisions is an order of magnitude less than their own self energy. That is because they must share that energy among all their quarks and gluons. Since electrons have no known substructure, they do not suffer from this handicap specific to composite particles. Collisions of electrons are therefore intrinsically more efficient and simpler than those of protons. The LHC project is the only one capable of energies high enough to explore the mechanism responsible for the mass of particles, now that the United States decided, on October 21, 1993, to cancel its own project, dubbed SSC (Superconducting Super Collider), which was even more ambitious and costly (in excess of 10 billion). At 7 TeV per beam, the energy of the LHC should be sufficient to test the existence of the Higgs boson. If it is discovered, the standard model would be strongly validated. If not, we can only hope that experiments conducted with this machine will suggest promising new directions for physics to explore.

Chapter 5

1. D'Alembert's "physico-mathematical sciences" included mechanics, geometrical astronomy, optics, acoustics, pneumatics, and the analysis of random events (a early form of statistics).
2. Jean d'Alembert, *Essai sur les éléments de la philosophie* (Essay on elements of philosophy), in *Oeuvres philosophiques, historiques et littéraires de d'Alembert* (D'Alembert's philosophical, historical, and literary writings) (Paris: Bastien, 1805), vol. 2, p. 367.
3. Michel Paty, *La Matière dérobée* (Matter stripped down) (Paris: Éditions des Archives Contemporaines, 1988), p. 45.
4. Edoardo Amaldi, "The Unity of Physics," in *Physics 50 Years Later*, ed. Sandorn C. Brown (Washington, D.C.: National Academy of Sciences, 1973), p. 21.
5. Michel Paty, *L'Analyse critique des sciences* (A critical analysis of science) (Paris: Éditions L'Hartmattan, 1990).
6. Julien Benda, *Exercices d'un enterré vif* (Exercises for someone buried alive) (Paris: Gallimard, 1946).
7. Cited in C. Chrétien, *La Science à l'oeuvre* (Science at work) (Paris: Hatier, 1991), p. 34.
8. Jacques Maritain, *The Dream of Descartes*, trans. Mabelle L. Andison (New

York: Philosophical Library, 1944), p. 28.
9. Max Planck, *Physikalische Abhandlungen und Vorträge* (Essays and lectures in physics) (Braunschweig: F. Vieweg, 1958).
10. Paul Feyerabend, *Against Method* (New York: Verso, 1988), p. 19.
11. Jean Ladrière, *Le Défi de la science et de la technologie aux cultures* (The challenge presented to cultures by science and technology) (Paris: Aubier-Unesco, 1977), p. 34.
12. Quoted in Arthur Koestler, *The Sleepwalkers: A History of Man's Changing Vision of the Universe* (New York: Grosset & Dunlap, 1963), p. 534.
13. Pierre Curie, "Sur la symétrie dans les phénomènes physiques, symétrie d'un champ électrique et d'un champ magnétique" (On symmetry in physical phenomena: The symmetry of electric and magnetic fields) *Journal de Physique Théorique at Appliquée* 3 (1894): 393-415.
14. Pierre Duhem, *The Aim and Structure of Physical Theory*, trans. Philip P. Wiener (New York: Atheneum, 1962).
15. A tax collector for the royal government.
16. Pierre Thuillier, *La Grande Implosion: Rapport sur l'effondrement de l'Occident* (The great implosion: A report on the collapse of the West) (Paris: Fayard, 1995).
17. Ernst Cassirer, *The Philosophy of Symbolic Forms*, vol. 3, *The Phenomenology of Knowledge*, trans. Ralph Manheim (New Haven, Conn.: Yale University Press, 1957).
18. Ibid.
19. Antoine Cournot was among the first to recognize, as early as 1870, that classical physics could not pronounce the ultimate word on the constitution of the world. Even before Gibbs and Boltzmann introduced the field of statistical mechanics, Cournot had foreseen the need for a new probabilistic description of nature: "While rational mechanics is one of the great paths that enable us to appreciate the economy of the world, there exists another one whose theory of combinations provides the key; it is a rougher path, perhaps less imposing and not as wide at first sight, but it also opens access in more varied directions; it was discovered, if not fully developed, in the seventeenth century." See Antoine Cournot, *Considérations sur la marche des idées et des événements dans les temps modernes* (Considerations on the march of ideas and events in modern times) (Paris: J. Vrin, 1975 [1872]), pp. 173-174.
20. The designation "condensed matter physics" is preferred over "solid state physics" because the field includes the study of liquids.
21. Roland Omnès, *Philosophie de la science contemporaine* (Philosophy of contemporary science) (Paris: Gallimard, 1994).
22. Benda, *Exercices d'un enterré vif.*
23. Steven Weinberg, "The Search for Unity: Notes for a History of Quantum Field Theory," *Deadalus,* Volume II (Fall 1977): 17-35. The cited comment appears on page 33.
24. Auguste Comte, *Cours de philosophie positive* (Lectures in positive philosophy) (Paris: Bachelier, 1830-1842), lesson 52. See note 33 to Chapter 2.

25. Gaston Bachelard, *The Philosophy of No: A Philosophy of the New Scientific Mind,* trans. G. C. Waterston (New York: Orion Press, 1968) pp. 8, 13.
26. Rudolf Carnap, "Logical Foundation of the Unity of Science," in *The Philosophy of Science,* ed. Richard Boyd, Philip Gasper, and J. D. Trout (Cambridge, Mass.: The MIT Press, 1991), p 397.
27. Michel Bitbol, *Mécanique quantique: Une Introduction philosophique* (Quantum mechanics: A philosophical introduction) (Paris: Flammarion, 1996).
28. Ernst Mach, *The Science of Mechanics: A Critical and Historical Account of Its Development,* trans. Thomas J. McCormack (La Salle, Ill.: Open Court Publishing Co., 1960).
29. Ernst Mach, *Antimetaphysische Bemerkungen* (Antimetaphysical remarks), in *Die Wiener moderne Literatur, Kunst und Musik zwischen 1890 und 1910* (Modern literature, art, and music between 1890 and 1910 in Vienna) (Stuttgart: Reclam, 1981), p. 145.
30. In one passage of his *Antimetaphysical Remarks,* Mach recounts how at age fifteen he happened to come across a copy of Kant's *Prolegomena to Any Future Metaphysics*. As he tells it, it was a revelation: "This work had on me a powerful and indelible effect, which no other philosophical book would ever match. Two or three years later, I understood the superfluous role played by 'the thing in itself.' While strolling outdoors on a bright summer afternoon, the world suddenly appeared to me as one single mass of interrelated impressions, embracing even myself. They just happened to be more strongly interrelated in me. Although I did not think anything more of it until much later, this was a defining moment for my overall view of things." The point is that it is the subject—and only the subject—that grasps the unity of the world. Mach also was the first physicist to proclaim that one cannot study a piece of the universe without considering the distribution of matter in its entirety. This idea was to be the cornerstone of Einstein's general relativity theory.
31. Jacques Roger, "Science et littérature à l'âge baroque" (Science and literature in the Baroque Age), in *Pour une histoire des sciences* (For a history of science) (Paris: Albin Michel, 1995).
32. Some see in the New Age a modern version of a triumphant holism, which views the world as an array of invisible networks connected to one another by subtle links, and in which innumerable resonances take place. In such a picture, nothing short of a superior and universal knowledge, inconsistent with specialization and compartmentalization, would be capable of grasping profound similarities and restoring the integrity of what is real. Unfortunately, by insisting on dealing with the whole as inseparable, one is condemned to a paralyzing inability to say anything about it. For what can be said about the whole? Strictly speaking, nothing at all. Evidence of the totality dispenses with the cumbersome and discursive processes of proofs and demonstrations since "it springs forth, like Minerva, all suited up in her armor, in a gratifying and outrageously simplifying meditation," as Michel Lacroix wrote in *La Spiritualité totalitaire, le New Age et les sectes* (Totalitarian spirituality: The New Age and sects) (Paris: Plon, 1995).

33. That is not to deny that the holistic approach has inspired some beautiful prose. What follows is an eloquent plea by Victor Hugo: "Nothing is truly small, as anyone knows who has peered into the secrets of Nature. Though philosophy may reach no final conclusion as to the original cause or ultimate extent, the contemplative mind is moved to ecstasy by this merging of forces into unity. Everything works upon everything else.
The science of mathematics applies to clouds; the radiance of starlight nourishes the rose; no thinker will dare say that the scent of hawthorn is valueless to the constellations. . . .The cheese-mite has its worth; the smallest is large and the largest is small; everything balances within the laws of necessity, a terrifying vision for the mind. Between living things and objects there is a miraculous relationship; within that inexhaustible compass, from the sun to the grub, there is no room for disdain; each thing needs every other thing. Light does not carry the scents of earth into the upper air without knowing what it is doing with them; darkness confers the essence of the stars upon the sleeping flowers. Every bird that flies carries a shred of the infinite in its claws." See Victor Hugo, *Les Misérables*, Part Four, Book III, The house in the rue Plummet, trans. Norman Denny (London: Penguin Books, 1982), p. 764.
34. Amaldi, *Physics 50 Years Later*, p. 20.
35. Pierre de Maupertuis, *Système de la nature* (System of nature) (Paris: J. Vrin, 1984).
36. Renée Bouveresse, *Karl Popper, ou le rationalisme critique* (Karl Popper: Critical rationalism) (Paris: J. Vrin, 1981).
37. This question is of interest to more than just the physical sciences. When pressed about the reductive character of his own thesis, the noted anthropologist René Girard answered: "On this point, I am in full agreement with Lévi-Strauss: Scientific inquiry is reductive or it is nothing at all." See René. Girard, *Things Hidden Since the Foundation of the World*, trans. Michael Metteer (Stanford, Calif.: Stanford University Press, 1987), p. 39.
38. P. A. M. Dirac, "Quantum Mechanics of Many-Electron Systems," *Proceedings of the Royal Society*, A123 (1929): 714-733.
39. A. J. Leggett, comment made during the question-and-answer session following S. Watanabe's paper "Is Reductionism Tenable within Physics?" in S. Kamefuchi, ed., "Foundations of Quantum Mechanics in the Light of New Technology," *Proceedings of the International Symposium on the Foundations of Quantum Mechanics (Tokyo: Physical Society of Japan, 1983)*, p. 264.
40. Paul Oppenheim and Hilary Putnam, "Unity of Science as a Working Hypothesis," in *Minnesota Studies in the Philosophy of Science*, vol. 2, *Concepts, Theories, and the Mind-Body Problem*, ed. H. Feigl, M. Scriven, and G. Maxwell (Minneapolis, Minn.: University of Minnesota Press, 1958), pp. 3-36.
41. Jean-Marc Lévy-Leblond, *L'Esprit de sel: Science, culture, politique* (The salty spirit: Science, culture, politics) (Paris: Fayard, 1981), p. 161.
42. The description of certain nuclear states involves explicitly the inner structure of nucleons (namely, quarks), as opposed to the properties of mere assemblages of nucleons.
43. Dominique Lecourt, *À quoi sert donc la philosophie? Des science le la nature aux*

sciences politiques (What is philosophy good for? From the sciences of nature to political science) (Paris: Presses Universitaires de France, 1993), p. 101.

44. Paty, *La Matière dérobée*, p. 51.
45. Ibid., p. 70.
46. François Dagognet, *Science et philosophie: Pour quoi faire?* (Science and philosophy: For what purpose?), ed. Roger-Pol Droit (Paris: Le Monde, 1990), p. 205.
47. Jacques Testart, *Des Grenouilles et des hommes: Conversations avec Jean Rostand* (On frogs and men: Conversations with Jean Rostand) (Paris: Stock, 1995).
48. Henri Poincaré, *Science and Hypothesis* (New York: Dover Publications, 1952), pp. 172-173. Poincaré felt strongly that science could progress only by swinging back and forth between two opposite poles, without either one winning over the other. In a 1912 conference talk entitled "New Concepts of Matter," Poincaré stated: "Science is condemned to vacillate endlessly from atomism to continuism, and from mechanism to dynamism. . . . This struggle [between antinomies] will last as long as science survives and humanity goes on thinking, because it is due to two opposite needs of man's spirit—the need to understand, and we can understand only what is finite, and the need to see, and all we can see are infinite stretches." See Henri Poincaré, "Les Conceptions nouvelles de la matière" (New concepts about matter), in *Le Matérialisme actuel* (Present-day materialism), Ouvrage Collectif (Paris: Flammarion, 1920), p. 53. The last statement is obviously eminently arguable. Poincaré further emphasized the need for scientists to judiciously pick and choose from among an infinite variety of facts those that are most likely to fit into a unifying picture:"In physics, the facts which give a large return are those which take their place in a very general law, because they enable us to foresee a very large number of others. . . . The only facts worthy of our attention are those which introduce order into this complexity and so make it accessible to us." See Henri Poincaré, *Science and Method*, trans. Francis Maitland (New York: Dover, 1952), pp. 29-30. No one can afford the luxury of examining all possible facts indiscriminately, because the nature of the human mind is such that it would get hopelessly lost. Hence the need for selectivity: "There is a hierarchy of facts. Some are without any positive bearing, and teach us nothing but themselves. The scientist who ascertains them learns nothing but facts, and becomes no better able to foresee new facts. . . . There are, on the other hand, facts that give a large return, each of which teaches us a new law. And since he is obliged to make a selection, it is to these latter facts that the scientist must devote himself." See ibid., pp. 284-285.
49. Erwin Schrödinger, *What Is Life?* (Garden City, N.Y.: Doubleday, 1956).
50. P. A. M. Dirac, "The Proton," *Nature* 126 (1930): 605.

Chapter 6

1. Henri Poincaré, *Science and Method*, trans. Francis Maitland (New York: Dover Publications, 1952), p. 26.
2. One is reminded of Leibniz's confession: "I thought I was entering the harbor, but...I

found myself thrown back into the open sea." See Gottfried W. Leibniz, *New System*, trans. R. S. Woolhouse (New York: Oxford University Press, 1997).

3. Friedrich Nietzsche, *The Gay Science*, trans. Walter Kaufman (New York: Vintage Books, 1974), pp. 335-336.
4. Erwin Schrödinger, *Mind and Matter* (Cambridge: Cambridge University Press, 1959), p. 67.
5. Arthur Schopenhauer, *The World as Will and Representation*, vol. 2, trans. E. F. J. Payne (New York: Dover Publications, 1969), pp. 172-173.
6. Physicists prefer to talk of "interactions," rather than "forces," between particles, since they not only affect the motion of particles, but also cause them to transform into one another. In classical physics, a force between two particles is transmitted via a field. The field created by one of the particles propagates through space and acts on the other (and vice versa). This concept had to be revised to make it consistent with the principles of quantum theory and relativity. Any interaction requires that "something" be exchanged. This "something" is what is called a quantum, or a particle characteristic of the field. In this picture, an interaction can take place between two particles only through the exchange of a third. This third particle, which transports the interaction, is called the gauge boson of the interaction. Because it cannot be detected directly, it is said to be "virtual" (which does not mean that it is not a real particle). The more massive the gauge boson, the shorter the range of the corresponding force.
7. The objects populating the universe make their presence known to us primarily through the visible light they emit. By extending the spectral range of detectors attached to modern telescopes, we have been able to discover many new objects in space, such as X-ray-emitting stars and infrared galaxies. By adding the mass of all these objects inferred from their luminosity, we can estimate the mass of the known universe, and from there its density, which plays a crucial role in cosmological models. There is nothing to guarantee that all possible objects can be inventoried this way. Many of them might emit no light at all, or too weak a radiation to be detected. The motion of stars or gases near the edge of rotating spiral galaxies seems to be determined by the mass of the galaxy, in accordance with Kepler's law, but with one important proviso: The mass of the galaxy has to be far higher than estimated on the basis of its luminosity.

   This stunning result had been suspected as early as the 1930s by the two astronomers Jan Oort and Fred Zwicky. It has been confirmed repeatedly since then with a variety of techniques. For instance, it is possible to measure the deflection of light rays skimming the gravitational field of a galaxy. The results indicate the presence of a huge halo of matter (as large as three hundred thousand light-years in radius) surrounding that galaxy. Dubbed "dark matter" because it emits no detectable radiation (the term has nothing to do with black body or with black holes), this mysterious substance would increase the mass of the universe by at least a factor of 10. No one understands the nature of that dark matter. Is it made of "brown dwarfs" (objects that are somewhere halfway between stars and planets and too dim to be seen), or are we dealing with the remnants of small stars called "white dwarfs"? The number of such objects detected so far is much too small to account for the huge missing mass. A number of physicists have suggested that dark matter involves some "exotic" (perhaps "supersymmetrical") substance left over from the primordial universe and whose detection appears extremely difficult.

Finally, the density of the universe can be estimated in one of two ways. Either it can be inferred by taking the ratio of the mass (luminous and hidden) and its presumed volume, or it can be calculated from cosmological models based on the measured rate of expansion of the universe. As it turns out, the two results are quite different. The total mass estimated with the first method is no more than one-fifth of the most likely value of the cosmological mass. The nature and location of the missing mass remains a complete mystery.

8. In the 1930s, physicists had already developed a theory of electromagnetism that included the principles of both quantum physics and relativity. It was called quantum electrodynamics. Unfortunately, it had a number of serious problems, not the least of which was a disturbing tendency to produce infinite values, totally unphysical, when called upon to calculate certain properties of atoms and particles. In time, physicists learned to get around that problem without really solving it. In 1947, Willis Lamb went on to measure an important physical quantity associated with the hydrogen atom, thereafter known as the "Lamb shift." Every attempt to calculate this quantity using quantum electrodynamics had until then led to infinities, which prompted many attempts to resolve this discrepancy between experiment and theory. After two years of intense efforts, theorists came up with a sophisticated mathematical technique that got rid of infinite quantities in the calculations and was able to predict results in remarkable agreement with experimental measurements. The procedure was dubbed "renormalization."

   Theories that are amenable to such a procedure, and which, additionally, involve only a finite number of parameters that can be determined experimentally, are said to be renormalizable. Only then can they be tested against experiments. Physical theories derived from a principle of local gauge invariance have been shown to always be renormalizable. Both quantum electrodynamics and quantum chromodynamics fall into that category.
9. Space and time are effectively dynamical variables in cosmology, much like position and velocity are for a particle.
10. Determinism in the sense of general relativity implies that knowing its present state is enough to know the complete model, including its entire time evolution.
11. The wave function assigns a probability to each possible state. At least, that is one possible interpretation. For instance, the spatial curvature of the universe may not be precisely defined, but the wave function provides all the necessary information about the distribution of possible values.
12. For the same reason, a trajectory cannot be defined in classical mechanics without knowing the initial values of the position and velocity.
13. On the downside, the traditional notion of time has to be discarded in favor of some "imaginary" time (in a mathematical sense), which makes its interpretation somewhat problematical.

Index

A Note on the Type

The Quest for Unity is set in Meridien and Frutiger. Both fonts were designed by Adrian Frutiger: The former in 1954 and the latter in 1975. Meridien was Frutiger's first text face; the eponymous font was originally designed for the signage at the Paris-Roissy Airport

Design and composition by Adam B. Bohannon.